Building Scientific

Gadgets

By Ron Newton

ISBN:1482368633

ISBN-13: 978-1482368635

Preface

It is the purpose of this book to provide the Scientist who can be a high school student, college student or a professional with inexpensive instruments that can be built in the home shop. Many of the instruments shown in this book cost thousands of dollars retail but can be built for less than $50.00 in parts. The instruments have been proven and published in several electronics magazines. The majority of instruments are physics oriented but also encompasses chemistry and biology. This book is compiled of both published items and other writings of the author.

As for equipment, at a minimal you will need a drill press. Access to a lathe and milling machine is a big asset. A list of tools is noted in the first chapter. Most people have a soldering iron and meter.

A special thanks to Dr. Earnest Ikenberry of La Verne University for starting me designing laboratory instruments.

R.N.

ABOUT THE AUTHOR

Ron Newton was born in 1938 and started electronics at the age of ten when his Dad gave him a set of headphones and a crystal set. His first degree was in chemistry as an analytical chemist. He past his boards in "Clinical Laboratory Science" in the 1960s and was the Director of several medical clinical laboratories. He worked his way up as a CEO of a small hospital. During that time, he also worked as a Bio-engineer. Offered a position as Director of Engineering at an Aerospace manufacturing plant, he earned another degree in the 1990s in electronics. He moved up as the Director of Research and Development. He has consulted for several universities and NASA.

He retired as Technical Director at a local Blood Bank in 2001.

He has been consulting ever since for the American Association of Blood Banks (AABB) under the auspices of the Center of Disease Control (CDC) working in Africa, Caribbean, former Soviet Union and South America, teaching Laboratory Techniques and Bio-engineering. He is also a writer for the Publication "Nuts and Volts" and holds several patents and he has over 130 inventions to his credit.

****

HOW TO USE THIS BOOK

Originally this book was designed to come with a CD in the back. However, the publisher I'm using does not have the ability to provide and insert the CDs. I have a created a website http://www.newtsbooks.com that you can go to and download the CD for free. This also has the advantage if errors are found I can update the files. Some of the projects use microprocessors and the software codes are located on the CD under the chapter name under "Program Files". To program the chip you will need a programmer or contact the author for pre-programmed chips. The software is written in assembly language with Microchip's® MPLab® which is free on Microchip's® website www.microchip.com.

Pictures, schematics and parts lists are also located on the CD to save on the cost of printing.

The board files and changeable printable schematics are also on the disk and can be opened by downloading Express PCB files which is free software from www.expresspcb.com. Don't worry; you won't be hassled by them. There are also"Hints and Tips" and source parts list on the disk. Most of the time I use Mouser or Digikey as suppliers. Clicking on the hyperlink in the Excel spreadsheet should take you there, however, about every three months I get a list of discontinued parts. If prices are shown in the spreadsheet, they will not be current as they are changing exponentially. 3 years ago a 1/8 watt resistor cost 2 cents and now it costs 9 cents.

Contents

NOTES

CHAPTER ONE

TOOLS FOR MAKING INSTRUMENTS

When in Africa, I often have to repair equipment and carry two types of tool kits. The tools I carry are listed on the CD. The tools listed in this chapter refer to the bench.

One of the tools I use the most is a Dataq Data Acquisition module. They sell for fewer than thirty dollars and are invaluable. They will record four channels plus two digital channels. The WINDAQ®/Lite software is free and is capable of performing filtering analysis the FFT and DFT functions. Or analyze any range of waveform data with its statistics functions. It can use X-Y plotting to examine the relationship of one channel to another. It will also do extended analysis functions allowing waveform peak detection, integration differentiation, arithmetic operations and more.

When you by a new laptop, save the old one and use it as a chart recorder. Several of the projects in this book need a continuous chart recorded for collecting data.

PROGRAMMERS:

Many of the projects use a Microchip® microprocessor. This processor must be programmed. The chips are also available from the author. If you wish to change the program or program the chips yourself, you will need a programmer. They are not expensive under $35.00. Take a look at PicKit 2. Having your own programmer opens up a whole new world to discover.

HELPFUL TOOLS IN THIS BOOK:

Nothing is worse than building a project, plugging it into the power supply and have it start smoking. Normally when this happens you have lost the only part for which you don't have a duplicate. There goes another week plus shipping costs of a dollar part.

One of the ways to prevent that is fuse the circuit. Take a look at Chapter two for simple device, or Chapter three which will also show you the amount of power being pulled and also will protect your circuit. Of course you can also use a 1/4 amp fuse and fuse holder in series with the power supply.

METERS:

You can purchase a good digital meter for less than $100.00. I personally don't like auto-range as they chase the voltage around and make reading confusing. Having two is a benefit.

I do a fair amount of temperature measuring with thermocouples and several meters have this capability. Also it is nice to be able to interface with a computer for continuous recording of voltage, ohms and amperage. Watch out! Jameco has several of these however, they are RS-232 and not USB.

SOLDERING EQUIPMENT:

I recommend a soldering station and not a cheap pencil. I use a Weller® 60 watt as I can change the tips for heavy to long conical tip 1/64" for surface mount. The tips also come in different temperatures but I consistently use 800 degree tips.

Use a small rosin core solder, .60mm (.025") 60/40 rosin core solder (Kester "44"). I'm a firm believer in QuickChip® "No Clean Tacky Solder Flux" for both soldering and de-soldering.

De-soldering stations are nice to have but most of the time I use Soder-Wick® Rosin size # 1 50-1-25 solder braid.

You can buy canned rosin remover, however it is nothing but alcohol. What I use is a paint spray can that you can fill yourself and charge it with compressed air.

Pick up some Horsehair bristle acid shop brushes and cut them to 1/4". They make a great rosin scrubbing brush.

PLIERS AND SCREWDRIVERS:

Don't sell Radio Shack to short. I have purchased German side cutters and paid up to $60 for them. However, my favorite nippers are a $7.00 pair from RS.

Buy the best comb pliers you can find as the cheap ones will only give you trouble. Use either 4" or 6" needle nose pliers.

For bench work, I buy a multi pack of micro screwdrivers and wrap red electrical tape around the cross heads for easy identification.

Use a 6" 1/4 & 3/8" slot screwdriver and a #2 & # 4 cross head.

USEFUL TOOLS BUT NOT NECESSARY:

OSCILLOSCOPE:

Bench top oscilloscopes are nice to have as they stay in one place; a dual channel is preferable for checking wave forms. You can get a decent one for under $500.00

The oscilloscope I carry for field use is a 5 megahertz Extech 381295. It is a dual channel. I have tried the Pico's which plugs into a laptop computer and turns the computer into an oscilloscope but found them to be unacceptable. Perhaps the newer ones are better.

FUNCTION GENERATOR:

This is a nice item to have available for generating sine, square and triangle waveforms.

POWER SUPPLY:

Great for building projects especially if they have current limiting.

SUBSTITUTION BOXES:

If you are going build your own projects both resistor and capacitor sub boxes are useful.

SUGGESTED HAND TOOL LIST

ELECTRONICS BENCH:

- Allen wrenches
- Heat gun
- Hemostats
- Jeweler's saw
- Jumpers micro and mini alligator
- Magnetic pickup
- Magnifier head mounted
- Marking pen
- Mirror inspection dental
- Moto tool
- Nut drivers
- Paint brush 1" (for cleaning)
- Pencil
- Razor blades
- Ruler
- Scalpel
- Shrink tubing
- Side cutters
- Scissors
- Solder
- Stripper
- Tape
- Triceps
- Velcro

- Wire Wrap Tool   RS 276-1570
- Wire
  Wire Wrap wire various colors   RS 278-501
  #22 solid and stranded various colors.
- Pliers
  - Slip joint
  - Needle nose
  - Diagonal
- Screw drivers
  - Jewelers
  - 1/4" slot
  - 3/8" slot
  - #2 cross
  - #4 cross

CHEMICALS:
- Acetone
- Alcohol
- Canned coolant
- Cleaner for pots
- Kriol
- Lacquer thinner
- MEK
- Turpentine
- WE 40
- Xylene

NUTS, BOLTS & SPACERS
- 2-56
- 4-40
- 6-32
- Nylon spacers # 4 & 6 various lengths.

MACHINE SHOP:
- Compressed air

- Drill press
- Drills numbered and 1/64 - 1/2" in 64ths
- Files, flat and Rattail
- Hacksaw
- Hammers
- Lathe*
- Milling machine*
- Power drill
- Taps 2-56, 4-40, 6-32
- Tin snips
- Vice
- Wrenches

*Expensive but useful only used in a few projects

TRAVEL TOOL KIT

CHAPTER TWO

DON'T BLOW A FUSE!! JUST RELAX!

Imagine yourself sitting on the veranda on the Serengeti in Africa. The sun is warm and you can feel the gentle breeze. From a distance, you can hear the elephants trumpeting and the faint sound of zebras running. On the roof there is a group of baboons playing and chattering.

All of a sudden there is the smell of burnt wiring. Your probe just slipped as your mind is not on troubleshooting.

This peaceful scene is interrupted by the sound of your cursing as you have blown your last fuse. The Serengeti is a vast place, filled with many animals and beautiful sites. A great place to see, but a difficult place to troubleshoot electronics.

As a retired clinical chemist, bio-engineer with a PhD in experience, one of the privileges I have is being sent all over the world with all expenses paid. I'm a teaching consultant for both government and non-government organizations, but I often act as a repair technician. (Tough job, but someone has to do it!) If you fix it, you are a hero. If you don't, no one seems to care as no one else has been able to fix it either. It's a win-win situation. And you meet a great number of nice people.

Unfortunately, there are not any part houses and TV repair places are non-existent in the areas I'm sent. One of the problems of my travels is carrying enough spare parts to complete repairs, especially fuses. They are difficult to obtain if not damn near impossible to find. I normally carry about two to three fuses of each amperage from .1 to 10 amps for types, 5 mms and the .25 inch standard fuse.

Many times a non-working piece of equipment will present itself with a blown fuse. This can be caused by a power surge and the varistor doing its job by blowing out the fuse. However, it is not uncommon for the power supply to be shot also. Ninety percent of the time when I'm troubleshooting, I don't have a schematic and it's often plug and replace with what you have in your box and see if the fuse blows again.

Within my bag of tricks, to avoid having to replace fuse after fuse thus depleting my precious supply, I designed a substitution box that uses circuit breakers. This box contains five circuit breakers 0.25, 0.5, 1.0, 5 and 10 amps. This seems to cover the most common types. I developed a fuse adapter which substitutes for the fuse. This has saved me a lot of grief and many fuses. I regularly use it in series with an amp meter to check on power consumption.

I also find it useful when prototyping at home as I place it in series with the power supply. I'm sure that most of you have placed a part in backwards only to see a puff of smoke coming from your project. Just use this small black box and you have peace of mind.

CONSTRUCTION:

This is one of the simplest projects you can build. I used a 2.4" x 4.4" x 1.4" hand box. The total cost is less than $35.00.

On the left side of the box I drilled three .625" holes for circuit breakers with two 3.23" holes in between for banana jacks. The holes are spaced .075" apart. One .323" hole is drilled on one end for another banana jack. This is the common connection (see Figure 1).

Each circuit breaker is wired to the common connection using a heavy buss wire or braided copper shield such as used for solder wick. The banana jack on the opposite side is connected to each different circuit breaker (see Photo 1).

The fuse adapters are made out of fuses with the glass removed. This can be done with a pair pliers or a vice. Place a piece of tape around the glass before crack the glass to prevent the glass from being scattered. Remove the excess glass from the ends of the fuse using a drill bit.

For the .25" x 1.25" fuse, I use 3/16" hollow PVC tubing available from most hobby shops. Cut the tubing to one inch and drill two 3/32" holes 1/4" apart from the center of the tubing. Thread two #20 gauge stranded wires into the holes and through the tubes out the ends. Solder the wires to the inside of the caps. Place some five minute epoxy glue into each end and gently pull the wires so that the PVC fits inside the caps.

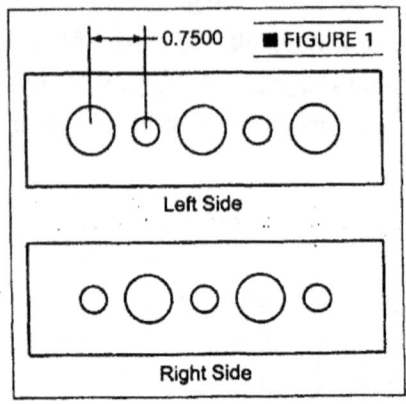

0.7500   ■ FIGURE 1

Left Side

Right Side

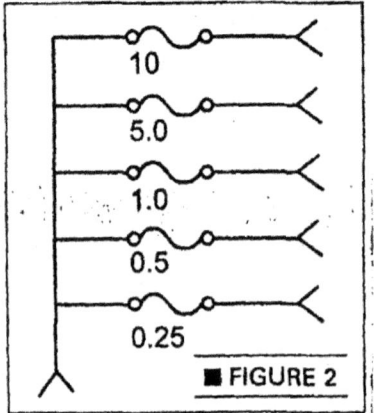

10

5.0

1.0

0.5

0.25

■ FIGURE 2

For the 5 mm x 20 mm fuses, I used the same 3/16" hollow PVC tubing but turned it on a lathe to 11/64" in diameter. You can use a drill press and sand paper if you don't have a lathe. Cut the tubing to 5/8" and follow the above procedure. Place two banana plugs on each end of the wires coming from the fuse adapters.

To use, remove the bad fuse, snap in the fuse adapter and plug one end into the common and the other end into the jack of the equivalent circuit breaker.   By adding a third wire with two banana jacks, you can place an amp meter in series with the fuse adapter.

Happy troubleshooting! As they say in Swahili, tutaonana and Asante.

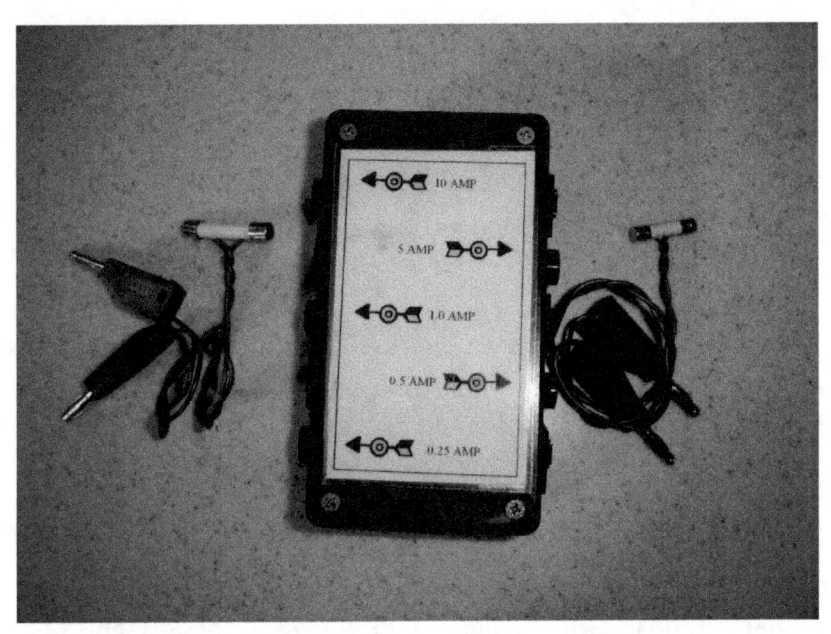

RESETTABLE FUSE BOX WITH LEADS

****

CHAPTER THREE

REFUSEABLE BOX
OR
JUST SAY NO TO TOO
MUCH POWER!

As some of you know from my previous writings, I spend a fair amount of time in Africa, mainly on the East coast, (Kenya, Rwanda, Mozambique and Tanzania), either teaching electronics or repairing used hospital and laboratory equipment. This presents some unusual challenges since there are no schematics and you don't have the advantage of just pulling a bad board and replacing it. The majority of the repairs are performed by substituting and replacing parts. To add to my frustration, the only country that speaks English is Kenya.

Learning equipment repair requires that the students have a very solid basic understanding of electronics. Recently we purchased several MPLAB PICkit 2 starter kits for teaching microprocessor programming. I developed a demo board kit which also teaches them soldering techniques for both surface mount and through hole components. This board provided 16 different programs with examples of tone generation, measuring temperature using linear sensors and thermistors, hall detection interface, photo cell, transducer interface and several others.

The students I teach have gone through a trade school, but have difficulty with a high a school pretest I give to access their needs.

Basic math skills are missing. Ask them to take 10% of 100 and they struggle. Calculate the resistance or voltage of parallel resistors and they just stare at you. The student's main complaint is that none of their teachers teach how to trouble shoot. This is most of what the students do once out of school. There is virtually no preventive maintenance done except in South Africa. Everything is making repairs in a panic mode.

What they have in lack in knowledge they make up in enthusiasm. They have a thirst for knowledge and are very bright people. One of my former students heard I was in Nairobi and came from Eldoret (200 miles) just to get advice on how to fix a baby's incubator.

Equipment is hard to come by in Africa, so one of my projects was to have them build a re-settable fuse box that shows the number of amps flowing when testing equipment while providing fuse protection. In Augusts' 2006 Nuts and Volts I published an article titled "Don't Blow a Fuse". The Africans could not obtain circuit breakers so my article didn't do much good over there. The parts for that project sold for about $40.00

In Africa fuses are rare commodities, so the ability to test without using one is most daunting. Since they have the PICkit 2 this seemed to be a logical step to provide another kit for them to build an updated version of a re-settable fuse box. The cost of this project came to $35.00 and does a lot more than the August 2006 version.

This project taught them of how to measure current using a voltage drop across a resistor, using a single supply amplifier op-amp configured as a non-inverting amplifier, working with seven segment multiplexed displays, and programming a Pic16F690 with an A/D and look up tables.

This circuit replaces an existing fuse with an inline relay that can be programmed to drop out using a pot for programming a range from 100 mA to 5 amps (256 steps). It displays the active ac or dc

13

amperage being drawn, and when the amperage of the circuit being tested exceeds the preset amperage the relay drops out.

I thought I would share this project as it is simple, easy and a very useful instrument to have for your tool box. If you don't have a PIC programmer, the chip and board are available at the end of the article.

Probably the most useful purpose of this device is to use it in series with a power supply with a new project you are building. This device can prevent burning out chips if you put them in backwards.

The unit is powered by a 9 volt battery and the voltage is reduced to 5 volts by a linear LM7805 regulator. A .01 ohm 1% tolerance current sense resistor is used to provide the voltage drop to be measured.

Using our old friend Ohms Law $E = IR$, if you have 1 amp flowing through a .01 ohm resistor the voltage across the resistor will be 10 mV.

The drop out relay is placed in series with the current sensing resistor. The voltage across the sense resistor is then amplified and sampled through an A/D. This amplifier is configured as a non-inverting amp with a gain of 50. Both ac and dc currents are passed through a 5 amp bridge. When measuring ac the diode provides full wave rectification. Dc passes through the bridge diode and therefore input polarity makes no difference.

The ac/dc switch puts in multiplication factors that corrects for minor differences.

One disadvantage of running the dc through the bridge diode is that a voltage below 1.6 volts will not pass due to the forward bias voltage of the diodes.

The unit is preprogrammed by holding down the program switch while adjusting the potentiometer to the desired setting. At that point the relay is held open. Once the program switch is released, the setting is recorded in memory of the microprocessor and the relay closes. Even if you bump the pot nothing will be changed unless the program switch is held down. Either ac or dc can be input. Pressing the ac/dc switch will alternate the input measurement mode. Once pushed and released the display will display the mode it is in for three seconds.

BUILDING THE UNIT:

The top side of the board has the components printing and the name of the board is on the bottom. Solder all the components to the board with exception **of the relay and the header**. The battery wires go through the hole on the bottom to act as a strain relief. (Make sure to put the battery holder in its compartment first.) The current resistor goes between the heavy traces on the top side. S1 and S2 use two 6 pin sockets to get height. Now place the relay along with the header on the bottom side of the board and solder.

ENCLOSURE:

Copy the template in Figure 1. Turn the top cover on its back with the spacer mounts facing away from you. Turn it over. Cut out the template and using glue stick paste it on the top of the box. *Make sure that rectangle cutout is facing away from you at the top...*

The enclosure is drilled using the template. The rectangle can be cut with a Moto tool or milled. The red lens is then cemented from the top side using superglue.

Two 5/16" holes are drilled into the half of the box with the battery compartment using the template. These two holes are for the mounting of the banana jacks in the box. The holes are spaced for ¾" male plugs.

S1 & S2 are placed into their respective sockets, and the micro is placed in its socket. You will need a programmer if you are going to use the code provided on the CD or you can order a preprogrammed chip from the author. If you have a PICkit 2 you can program the micro by using the header.

Solder two 1" green wires to the bridge diode ac input. Solder a 2" black wire to the minus terminal and a 2" red wire to the plus terminal. Use shrink tubing to insulate the solder joint. Glue or screw mounts the diode to the bottom of the box. Solder the two green wires to the jacks. Solder the black wire to the – hole on the board and the red to the + hole on the board.

Once the enclosure is prepared, the board can be slipped into the top side of the enclosure and mounted using 2 # 4 3/8" self tapping screws.

FUSE ADAPTERS:

The fuse adapters are made out of fuses with the glass removed. This can be done with a pair of pliers or a vice. Place a piece of tape around the glass before cracking the glass to prevent the glass from being scattered. Remove the excess glass from the ends of the fuses using a drill bit.

For the .25" x 1.25", I used 3/16" hollow PVC tubing available from most hobby shops. Cut the tubing to 1 inch and drill two 3/32" holes ¼" apart from the center of the tubing. Thread two # 20 gauge stranded wires into the holes and through the tubes out the ends. Solder the wires to the inside of the caps. Place some 5 minute epoxy glue into each end and gently pull the wires so that the PVC fits inside the caps.

For the 5 mm X 20mm fuses I used the same 3/16" hollow PVC tubing but turned it on a lathe to 11/64" in diameter. You can use a drill press and sand paper if you don't have a lathe. Cut the tubing to 5/8" and follow the above procedure. Place two banana plugs on each end of the wires coming from the fuse adapters.   See Figure 2

For testing your projects, use two flexible wires with alligator clips on the ends.

HOW TO USE IT:

1. MAKE SURE THE POWER IS OFF ON THE PIECE OF EQUIPMENT YOU ARE GOING TO TEST.
2. Remove the fuse and replace it with one of the jumpers.
3. Plug the jumpers in to the "Refuseable box". If testing one of your projects, make sure the polarity is correct.
4. Turn on the "Refuseable Box" by pressing the on/off switch.
5. Prg will be displayed. Hold down the "Program Switch" and adjust the pot for the amperage tripping point you want. Release the "Program Switch"
6. The unit will display "AC" indicating that you are in the ac mode. *(During this period the relay will be open for safety reasons.)*
7. If you want the dc mode, press the "AC/DC" switch.
8. Now turn on the piece of equipment you are going to test.
   a. If the equipment exceeds the preset amperage value the relay will open and the unit will display "BAd". **Turn off the equipment you are testing!!** *You will have to turn off the "Refuseable Box" to reset the relay.*
   b. If the power is less than the preset value it will display the amount of amps it is pulling.

Happy are those who just RE-FUSE!

REFUSEABLE FUSE BOX

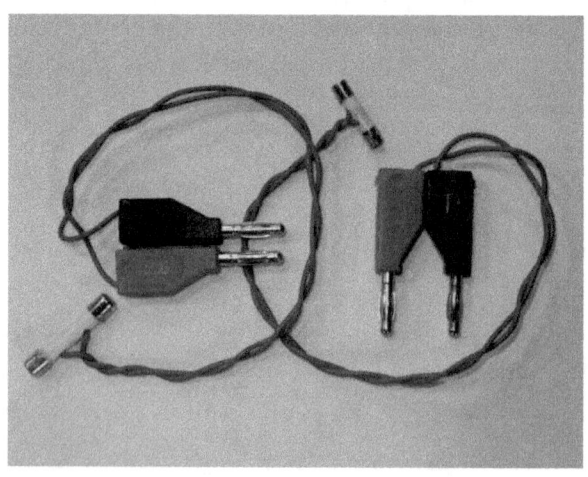

CHAPTER FOUR

KING TUT'S PYRAMID GRAVIMETER

August 15, 2012 Published in January 2013 Nuts and Volts

Gravity is that mysterious force that most of us fight when we step on the scales. Actually it is not our weight we need to worry about but our mass. If it wasn't for ancestor Isaac Newton, we would all be floating around space as he discovered gravity. Some people believe that the pyramids were built by aliens using anti- gravity technique. The weight is in the shape of a pyramid, the title is with tongue in cheek; however this is a very serious working instrument. I only wish I had it when we had the total eclipse this year as there are many unanswered questions regarding gravity when the moon blocks the sun. If anyone is going see the eclipse in November 3, 2013 in Australia, let me know.

With the recent discovery of the God particle it probably puts us closer to the proof if the graviton particle exists. Gravity is different in areas of the earth for several reasons. On high mountains you weigh less due to being further from the mass of the earth. At the equator, centrifugal force plays a part versus measurements at the north and south poles. A one ton object weighs 7 lbs less at the North Pole than at the equator. I think you will be amazed at the number of variations you will see in gravity.

Super novas and black holes also affect the earth's gravity. These can also be measured with this Gravimeter. It has been reported that gravity changed from 1% deviation to 2% deviation in the 1980s. The Gravimeter also has practical applications in discovery of large deposits of oil and minerals as their mass differ. Oil is often found under salt domes which have less mass than that of iron. So if you are looking for buried treasure, here is another gadget for your tool box.

Commercial gravimeters can be very expensive. They often employ a pendulum method which was used by Henry Cavendish in the late 1800s. Balance Scale Gravimeters are also used using a spring and a weight. I tried several types of sensors and decided on using a very simple one detecting the change of resistance by a weight using conductive foam. This is equivalent to a spring type of scale gravimeter. It has an advantage of being rugged as there is no spring bouncing up and down. However it has a disadvantage with time with the foam aging and losing its elasticity properties.

This project can be built for about $30.00 and is easy to construct as it has no surface mounts. It uses an inexpensive LCD display that does graphics and alpha numeric characters all driven by 11 inputs.

MULTIPURPOSE BOARD:
 Studying the ASM code of the PIC 16F916 included on the CD will give you an idea of how to program the display for other projects. I have included programming pads for the use of a Microchip Pic 2 programmer. The board can be used for other projects e.g., measuring resistance, capacitance, voltage or just plain counting events by changing the code. I have added extra pads for those who wish to use the display board for other purposes.

ELECTRONICS:

- The sensor is a variable resistor comprised of electro conductive foam. It changes resistances when squeezed together. It has a resistance of approximately 1,000 ohms when constructed. The Pic microprocessor uses a process for calculating the time it takes the resistor to charge a

capacitor. The formula for charging a capacitor is R * C = t.
R is in ohms and C in Farads. "t" is in seconds.

$$1,000 * 100uF = .1 \; seconds$$

The TMR0 register is used as it advances one count every .000, 001
of a second. When it rolls over to 255 it causes an interrupt and
increments two more registers until the voltage exceeds a preset
amount in the comparator. 1,000,000 per second x .1 seconds =
100,000 counts. Since gravity can vary ± 2% the count can vary
2,000 counts increasing its accuracy. There are a number
algorithms built into the software which perform addition,
subtraction, multiplication and division. Once the base count is
established the difference in counts between the base count and
new count is determined. This difference is divided by the base
count time times 100, will give a % change of the gravity field.

$$\frac{Difference \; between \; New \; count \; and \; Base \; Count}{Base \; count} * 100$$
$$= \% \; change \; of \; gravity$$

The Pic Microprocessor used also has memory areas to store
previous data. When pressing the zero switch this zeros the base
count number and stores it in the EEPROM area. Pushing the zero
button while pressing the on off switch will load the base count. If
not pushed, it will calculate a new base count after averaging 25
counts.

The unit is powered by a 9 volt battery. IC3 is a LM7805 5 volt
voltage regulator. The circuit draws 3.5 ma and the battery should
last about 6 days. C1 smoothes out any ripples. If you are going
leave it on all the time, use a 9 volt battery eliminator.

The microprocessor shorts outs the charging capacitor C2 and then
voltage is applied across the sensor. The capacitor will charge and
when its voltage reaches a preset voltage of the comparator it

signals the micro to calculate the number of counts the TRM0 register it has made during this charging period. The math functions take over from there. The display displays to three decimal places. The DAC is a 12 bit digital to analog converter. Its voltage is centered at 2.5 volts for a zero position on chart recorders. The ± % change is added or subtracted from 2.5 volts. This gives a very sensitive graph. My first recording went off the chart so I reduced the chart reading by 2. ∴ ± 1 volt from 2.5 volts = 2% gravity change.

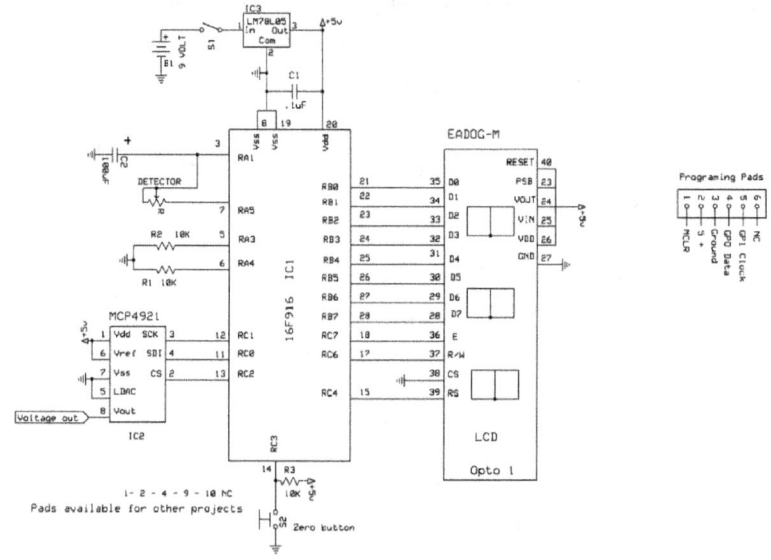

MODIFICATION OF THE CHASSIS:

The first thing to do is put the box together with screws as they have to form threads. This will make it easier to assemble when you put everything together. Go to the CD and download the chassis templates. Cut the template and using glue stick paste on the top of box. Drill the five holes as called for on the drawing. The templates can be removed using hot water.

Place the jack from the left outside of the box and secure it with its nut.

THE DETECTOR:

Cut two 1.25" square single sided copper boards. Solder two 4" wire wrap wires to each plate. Cut the wire off the bottom 2 oz pyramid sinker, sand flush and glue to the top pad on the non-conductive side. **Do not use** steel weights as they will be subject to magnetic fields.

Using double backed tape, press the other copper board copper face up to the bottom of the chassis center about 1/8" from the side opposite the battery compartment. The wire should be facing the battery compartment. Cut a 1" square conductive 1/4" foam that electronic parts are shipped in. Put a small drop of Graphite-Filled Conductive Wire Glue Radio Shack # 64-146 on the bottom corners 1 & 3 of the conductive foam and place on the bottom copper board. Add a drop of glue on corners 2 & 4 on the top side of the foam and center the other board. Allow to dry over night. Do not use too much glue as it will flow through the foam and short it out. You must use conductive glue. Treat the sensor gently as the foam will tear from the boards. The sinker will help hold the sensor while it sets.

After dry, using an ohmmeter, check the resistance of the two plates, it should be around 500 to 1,000 ohms.

CIRCUIT BOARD:

The circuit board files are generated using Express PCB and are on the CD.

Use a 6-32 tap and tap the 2 corner holes.

Solder R1, R2, R3 and C1. Note the polarity of C2 and solder the positive terminal to the square pad. Solder IC 1 and IC 2 noting pin one goes to the square pads. Solder IC3 with its flat pointing toward the edge of the switch. Solder in Opto 1. Solder the on - off switch. Turn over the board and solder momentary switch to the

23

bottom side. If you purchase a pre-programmed chip skip the next paragraph.

The chip can be programmed using the 5 program pads under the LCD. Use a Pic 2 or a Pic 3 programmer. The assembly file is on the website.

Using two different 4" colored wire wrap wires to the pads next to pin 7 and pin 8 of the DAC. The wire from pin 8 goes to the center post of the RCA jack and pin 7 goes to the ground. Pass the two wires from the 9 volt battery holder through a hole in the battery compartment, through the strain hole and solder the red wire to the + terminal and black to the - pad. Solder the two wires coming from the detector to the "in" pads. They have no polarity.

Place two 1" 6-32 screws in the two holes from the top side of the chassis. Turnover and add the 3/8" standoff to the screw opposite the switch. Screw the board to the top of the chassis. The tapped holes will act as nuts. When finished, add a drop of finger nail polish to the ends of the screws to prevent them from turning.

USING THE UNIT:

Add a 9 volt battery. If using a data logging software, plug the data acquisition module into the jack.

Place the unit on a level flat surface and push the on switch. The unit will display "LOADING". Push the zero button, the display should change and display "STANDBY". After about 30 seconds the display will change to "STORING" and then display percentage. You can zero the instrument any time by pushing the zero button on the side. It will store the base readings in the EEPROM. Keep in mind that the unit will be subject to temperature change and probably barometric pressure due to the foam bubbles. It also makes a good seismic detector.

Gravigram:

I have several chart recorders that I use throughout the years. They use reams of expensive paper. I have replaced them with my old computers and data acquisition units with software. Everyone ends up every few years replacing their old computer. They make ideal chart recorders. Dataq (www.dataq.com) makes a data acquisition module which sells for under $30.00 and measures 4 analog signals along with two digital inputs. This is a must for any scientist's laboratory. I have been using several of their products for 20 years or more. You just connect the module to your computer and it charts your data. With their free software you can compress a month of data into one page for printing.

The DAC produces 0-5 volts. Connect the data acquisition unit by using a RCA plug. Plug it into your computer and call up the software. There are a plethora of commands just refer to the software manual. It will monitor on the screen a live display of the voltages. It will sample up to 240 Hz giving you a magnetogram. I have used these types of modules for seismographs recording, recording a month of data at a time. You can save the data and call it back looking at individual periods of one minute. For smoothing of the data getting rid of the rapid changes of gravity, check the CD and download "Hints and Tips".

If you don't want to tie up a computer and for field use or stand alone, Dataq makes voltage monitor that plugs into a USB port and charts the results. The EL-USB-3 USB Data Logger sells for $75.00 (www.dataq.com click on USB data loggers). It will sample from one second for 9 hours or every minute for 22 days or for other applications sample 12 hours for > 2 years.

Start the EL-USB-3 USB Data software and set its parameters. You can use an immediate start or a delayed start. It even has alarm led's. Connect it to the red and white wires noting the polarity. You can remove it at anytime and download its data using the USB port and it will chart your data for you.

To determine the gravity of a situation, you must weigh all the facts!

KING TUT'S GRAVIMETER

****

CHAPTER FIVE

A POOR MAN'S SEISMOGRAPH

By Ron Newton

June 21, 2011

Now that Washington D.C. knows that they are not exempt to earthquakes, I have been receiving e-mails about a former article I wrote on building a seismograph.

I have been using seismographs most of my life and I have three in the back garage. I published several articles on seismic sensing and though it was about time to update to the latest chips, software and the use of SD memory chips. For those that are interested in previous articles, see "Measure the Earth Tides with a Tiltometer", *Electronics Now*, May 2003; "Seismic Detector" *Electronics Now* November 1999; and "Vibration Detector" *Nuts & Volts* 2010.

This is a great project for Science Fairs and those who wish to do their first soldering project. There are only a couple of surface mount 805 resistors to try your hand at surface mounts.   If you wish to have your own boards made, I have included on the CD, the Express PCB board files the Microchip ASM files along with Hints and Tips. Express PCB provides free software for board design and schematic design. www.expresspcb.com .

When I started writing this article, I thought I would use the standard Geo Phone that back in 1999 you could buy used for $8.00. Now they are on e-bay for over $60.00, too much for the hobbyist or too pricy for the High school student and difficult to obtain. I put on my thinking cap and went back to the old pendulum method, one of the methods still being used for seismology. The word 'pendulum' comes from the Latin word *pendulus*, meaning 'hanging'.

The period of the pendulum is determined by the length of its swing and not its weight and is given by the formula of $T \approx 2 * \pi * \sqrt{L/g}$ .Where T is in seconds, L is in meters and g is the acceleration of gravity. I decided to use a pendulum with a length of 9.8" (.249 meters) to give a period of 1 second.

$$1 \sec \approx 6.28 * \sqrt{\frac{.249 \ meters}{9.8 \ meters/second^2}}$$

However, the weight of the pendulum does have purpose as it has to keep the bottom of pendulum steady while the seismic activity moves the top end of the pendulum.

The sensor used is a piezo electric film that generates a small voltage when bent. The brass bar adds the mass and the piezo sensor acts as a damper. The voltage from the sensor is amplified by an operational amplifier that feeds into a microprocessor.

When starting a project, I always put down the starting specifications I want to accomplish which are subject to change. In this project, this is what I started with:

1. Inexpensive
2. Portable
3. Water proof
4. Battery operated with a battery life of 1 year
5. Indicator when seismic action has taken place
6. Data to be removed and downloaded into a computer

The cost of building the instrument is below $35.00 dollars.

The container for the seismograph is made out of a 13.5" 3" O.D. PVC tube with either an end cap on the bottom or a stand using a 3" flange. A piece of Plexiglas on the top is used for viewing the led to see if seismic action has taken place. This container can also be made waterproof.

I ended up with using 2 C cells for the power supply. It was determined that the unit draws 1 .5 ma when in standby and the C cells provide six months of service. The series J microprocessors draws less than .1uA when in standby, however when the comparator is used it draws power along with the op-amp which increases its standby power. The SC chip takes most of the power. The C cells sit in the bottom of the PVC tube.

The seismic detector is made out of a Measurement Specialties Piezo film vibration monitor. As the piezo film is displaced from the mechanical neutral axis, bending creates very high strain with the piezopolmer and therefore high voltages are generated. A piece of brass bar measuring 9.8 inches is added to add mass using a #1 screw and nut. This bar acts as a pendulum. The sensor provides about one second of damping.

Electronics:

The Piezo film sensor generates a small amount of voltage when bent or stressed. This voltage is amplified by the use of a FET non-inverting operational amplifier. The amplifier in this configuration gives a very high impedance input. A Microchip MCP 601 was used, as it can be run using a single voltage supply. The voltage is injected in the non-inverting port configured as an adder summing a voltage provided by a voltage divider potentiometer R4. This voltage is amplified by 11 times with resistors R1 and R2. The amplification of a non-inverting op amp is:

$$amplification = 1 + \frac{feed\ back\ resistor}{grounding\ resistor}$$
$$11 = 1 + \frac{100k}{10k}$$

The vibration waves fluctuate above and below a 1.6-volt offset provided by R4. A 10 megohms resistor prevents offset loading to the transducer. The offset allows ± voltages of the sensor to be viewed on a chart without using ± power supplies. The micro measures the voltage 120 times per second.

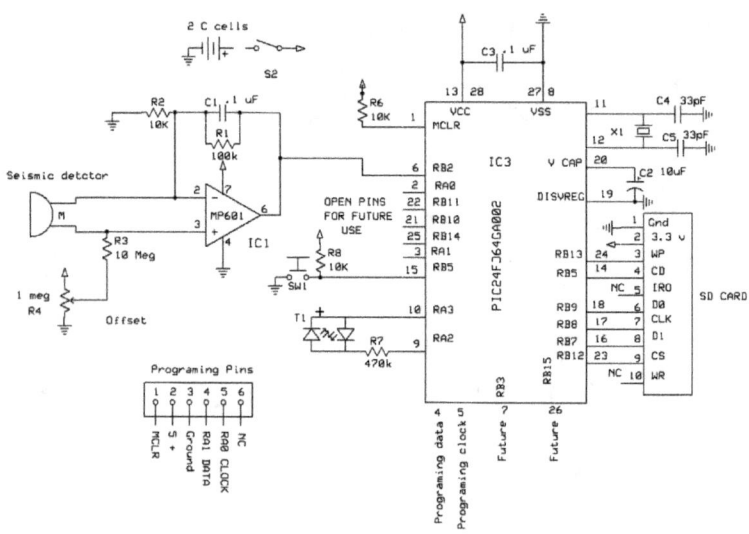

The bi-colored led is turned on /off and changes its color by applying opposing voltages to its leads. The led can generate three colors, red, green and yellow. It can also be made to flash.

The microprocessor takes the amplified voltage and is channeled to the micro's comparator, which continues to run when the micro is put to sleep, thus the micro draws very little power. When the voltage of the sensor is compared against a 1.6-volt voltage reference and is exceeded, the micro switches from the comparator mode to the analog to digital converter mode measuring the voltage digitally. The micro opens the first text file on the SD card that is named Seismic1.txt. It then will record one minute of data, and then flash the led every 4 seconds indicating that a seismic event has taken place. It then waits for another earthquake and the process starts over again, however, the text file is incremented to Seismic2.txt and will continue endlessly. The unit can be downloaded at any time by turning off the power switch.

The SD card is then removed and placed into a computer for downloading and viewing.

A Microchip PIC24FJ64GA002 was used for this project as there is plenty of documentation on how to interface with SD cards. (See N&V October 2010 "Implementing a File I/O for the 16 bit Micro Experimenter" by Thomas Kibalo. The Experimenter kit is also available on the N&V online store.)

CONSTRUCTING THE BOARD:

There are two boards for this project. The first one is for the SD card.

Turn the board over and solder the six 10K 805 surface mount pull up resistors. The best way of doing this is to melt a small amount of solder to top six pads. Using tweezers to hold the 805, touch the tip of the soldering iron to the pad and place the 805 on the pad allowing the solder to flow to the resistor. Do this for rest of the resistors. Now using solder, solder the other side of the resistor to the pads. Make sure you use rosin core solder. I normally use .6 mm diameter. Turn over the board and solder the SD holder to the top side. Solder each of the pins and ears in their respective holes.

The second board will need to be trimmed to the outside circle. The easiest way is to use a sander. Make sure it will slip into the 3" PVC pipe so you don't have to sand after the components are placed on the board. All the components are placed on the top of the board with the exception of the sensor.

Solder IC1 and IC2 to the board. Note the square hole is pin one of the chips. Place the chip in front of you with its notch pointing left. Pin one is the lower left pin. Often there is a small dimple above it. Watch IC2 as pin one is reverse that of IC1. Solder The 1 Meg pot,

resistors, capacitors and the xtal. Note the polarity of the 10uF cap.
Solder the two switches. Solder the led with its long lead going to
the square pad. Cut two headers, one with 10 pins and the other
with five pins. Solder them to the board. There will be extra pads
next to the microprocessor that can be used for other projects.
Nothing will go in IC3 as this is for a future project. There also will
be empty pads for those who program and wish to add alarms or
other items.

Take the SD circuit board and place it over the 10-pin header. It
should rest on micro. Solder the 10 headers

Cut the ¼" x 1/16" brass bar to nine inches. Drill a 1/16" hole 1/8"
from one end of the brass bar. Use a #0- 80 3/8" screw and nut and
fasten it to the Piezo sensor. Add a drop of fingernail polish to the
nut. Solder this assembly to the bottom of the board.

Solder two 13" wires to the battery holder. Make sure you can
identify the positive wire.

For those who bought a pre-programmed chip, skip this paragraph.
The chip can be programmed or re-programmed repeatedly using a
Microchip PicKit II. The files are zipped and are on the CD. The
programming was written in C. The square pad is Pin one for the
programmer. There are only five pins needed (the sixth is
redundant).

CONTAINER:

There are two types of containers that you can use for this project.
One is waterproof for placing in the ground and the other is a
standalone, which can be mounted on the floor.
The container is made out of a 13.5" piece of 3" PVC pipe plus a 3"
flange for standalone or an end cap for burying. The top of the pipe
should be smooth.   The PVC pipe will be loose in the flange. Having
a lathe, I turned ½ of a 3" collar to fit the flange. Another way is to
use duct tape and build up the pipe to fit the flange.

Go to the CD and download the files which go to this project. There
you will find a template for the Plexiglas lens. Cut out and using a

glue stick glue it to a ¼" piece of Plexiglas. Sand the Plexiglas to the edge of the circle. Using a 9/64 drill, drill three holes. Remove the template using hot water. Place the lens over the smoothed end of the PVC pipe. Mark the pipe where the holes are. Using a #36 drill, drill the three holes ½" deep. Tap each hole with a 6-32 tap. Using three 1" screws add a drop of super glue to each hole and screw each screw into the PVC pipe. Use a cut off tool and remove the heads. If using a flange and not a cap, use the bottom template to cover the bottom. Glue this piece to the bottom.

Using the seismograph either inside or out you will need the lens to protect against air movements. If you are going to use the unit outside and if needs to be water proof, buy a 3" "Flush Valve Washer" and using the lens, mark the gasket and punch, three holes. Place the gasket over the screws. If done properly the lens should fit over the screws and be secured by three knurled nuts.

Cut a piece of .06" x .125" Styrene plastic strip to 7.75 inches. Bend the strip and place 5/8" from the edge of the PVC pipe. Using super glue, glue this strip in place. This will act as a support for the board.

If using an end cap, place the PVC pipe into a bucket or a coffee can to keep it vertical. Load the battery holder with 2 C batteries and slide it into the bottom of the PVC pipe. Use small pieces of tape to hold the wires to prevent them from hitting the pendulum. Slide board and its pendulum into the tube and feed the two wires though the hole next to the pot. Try not to let them short. Allow the board to sit on its support. Place the two wires in their respective terminals and tighten. You might want to mark the positive terminal with a drop of red fingernail polish. **DON'T REVERSE THE LEADS!**

Turn on the power switch. The led should turn red indicating that power is being supplied. Turn off the unit.

 Hold down the red switch and turn on the unit. This will place it in the calibration mode. Release the red switch. If the led is red, turn the pot clockwise until it turns yellow-orange. If the led is Green, turn the pot counter clockwise until it turns yellow-orange. Set the pot when the led just turns red. This sets the trip point. Turn off the unit. If the unit is too sensitive, turn the pot CCW a 1/8 of a turn.

USING THE UNIT:

Using your computer*, format a SD card from 256 MB to up to 2.0 GB. A seismic recording only takes up 35.1 kb. That's many earthquake recordings.   Now we are going to set the time in the microprocessor's clock. **NOTE: If the power is turned off, the micro's, clock will be lost.** Download the Excel sheet titled "Time Calculations for Seismic". Follow the directions on the spreadsheet. The time is written to the SD card using a BCD format in hexadecimal code. Four six bytes of code in one long statement are stored in time.txt.   Make sure you allow enough time to place the card into the unit.

Load the card. Turn on the unit. The led should be red. Watch the clock and when it reaches the proper minute, press switch one and release. The led should flash green indicating that the micro's clock has been set to the time you programmed on the SD card.

Once you have located the place you are going place the seismograph, you can activate it by pushing the switch. The led will turn on steady red for one minute. This will allow you to place the lens on and secure it with the knurled nuts. When the red led goes out, it indicates that it is armed.

When the unit is jarred or an earthquake happens, the data will be written to the SD card and the led will turn a yellow – orange indicating that it is making a measurement.   Once the unit has made a one-minute measurement, the led will flash red about every 4 seconds indicating that an earthquake has occurred.

When you are ready to read the data, remove the lens (this will activate the unit and record the jarring of the removal also but will end up as an invalid txt file). Turn off the power and remove the SD chip.

VIEWING THE DATA:

To view the data, go to www.dataq.com and download their free software under downloads titled "WinDaq Waveform Browser ". For those that are interested in great data loggers and data acquisition I can't say enough good things about "WinDaq". The browser is also located on CD.

Once the"Waveform Browser" is downloaded, there should be an Icon on your computer. Click this icon. A window should open up. Locate the SD card and make sure you have "Files of type" on "(*.txt)", look for "Seismic1.txt" file or other numbered seismic files. Click on this. Perform a save Seismic1.wdc (default). Use #2 "Spreadsheet print file (ASCII)". The convert screen should pop up and default on "volts". Hit the return key. (Subject to change depending what version, you have with the Browser.)

The Waveform Browser should now be visible. If you go to "options" and click on "Cursor time", the date and time will show at the bottom of the time the earthquake happened. Moving the cursor below the red line, will show the time. By pressing F4 to place a time marker you can measure from the time marker the seconds from an event.

There is a plethora of commands you can use in the "WinDaq" browser, and I would defer you to their help menu, as it would take a small book to explain them all. Once you have the screen up you can calculate the "S" & "P" waves and their times to determine how far the epicenter was located. Keep in mind that it takes at least three seismographs to locate the epicenter with triangulation. Once you have down loaded your data, delete all the seismic.txt files on the SD card to re-record.

Shake Rattle and Roll!

Note: Recently a reader wrote me and had found Geophones for $10.00 ea on e-bay. This will reduce the size to three inches.

SEISMOGRAPH

CHAPTER SIX

THE ELECTRONIC SNIFFER

My wife and I were taking our morning constitutional walk up the hill, when I noticed the smell. The wind was from the South. "Do you smell that?" "What?" she said, "I don't smell anything" "Our neighbor must be gluing some PVC pipe or fiber glassing something?" I didn't think much more about it, but the smell continued off and on for the next couple of weeks.

A month later the wind was from the North and we had taken a different route. "Wow!" I said, "Is that ever strong!" "I don't smell a thing" she replied. It then hit me like a ton of bricks. I am a retired chemist and I don't know why it hadn't clicked before. Acetone or ether or a combination thereof. There was a drug lab in the neighborhood!

I went home and scrounged through my junk box and I found a Figaro TGS 822 gas sensor. I bread boarded it up to a 12-volt battery and hooked up a voltmeter. It still functioned. Our noses soon become desensitized to odors. The Figaro gas sensors do not. It didn't take me long walking around the neighborhood to find the house the smell was coming from. I called drug enforcement and explained my suspicions. Three months later there was a drug bust.

Everyone does not need one of these things, but it has many other uses. By changing the sensor you can detect the following. Propane, methane, hydrogen, carbon monoxide ammonia, hydrogen sulfide, organic solvents, CFCs and carbon dioxide to name a few. Methane is found in mines and is odorless. Miners used to carry a canary to detect it. So if you are exploring mines it might have a use if you don't have a canary. Carbon monoxide is also odorless and is a deadly killer and can be found in tunnels. Ever wonder what the carbon monoxide level is in the Holland Tunnel? The Carbon monoxide unit can be used for checking out bad gas heaters or has use with fire departments. The TGS 822 is good for organic solvents.

The project presented is simple. It uses a round board that fits into a 4-D cell Mag-lite available at any hardware store. You can find them for less than $20.00 if you shop around. The good thing about using this flashlight as a power source is there are no modifications required and you are able to still use it a flashlight. Many of the sensors like the TGS 822 have a heater and draw a hefty current 660 milliwatts or 132 milliamps. The D cells will power the unit for 24 hours. The TGS 822 cost is $14.50 as of the January 2004 price list. Other sensors vary in price. All are obtainable directly from Figaro. (Also see Parallax website.)

METHOLOGY:

The heater requires a 5 volt regulated power supply. I used a National low drop out L4931CZ50-AP, which will allow operation down to 5.1 volts but don't exceed 7 volts input. A Wheatstone bridge is used to provide the most sensitivity and corrects for the ambient temperature. As the sensor changes its resistance when it encounters a gas, the voltage changes across the Wheatstone bridge. The brain of the unit is a PIC12C671 A/D microprocessor. The voltage from the Wheatstone bridge is fed into the microprocessor and with some simple math algorithms is converted into sound. A Bi-color led is used to make sure you are on the right side of the cross over point of the Wheatstone bridge.

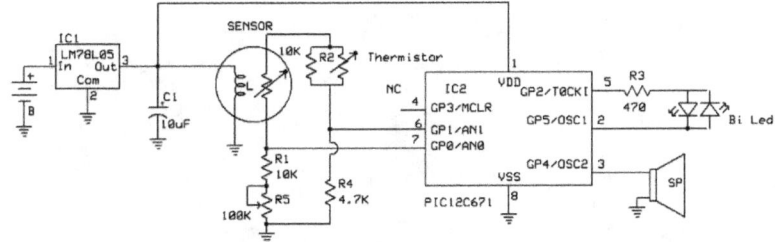

CONSTRUCTION:

You will need a programmer to program the PIC chip or it is available pre-programmed. (See parts list.) The code is listed on the CD.

All of the components with the exception of the terminal and the voltage regulator are soldered to the top of the board with the word "Sniff". The sensor's six pins are designed to be mounted regardless of it position; however, the sound transducer is marked with polarity. The long pin of the led goes to the pad marked +. There are 4 holes for different pin spacing of different sound transducers.

The voltage regulator is soldered to the bottom side of the board along with the 2-pin terminal. Pin one is marked for the voltage regulator. Take some red fingernail polish and mark the terminal next to the R (red wire) for ease of wire connection.

To make the bulb adapter, use a PR4 extra bulb and break the glass. Clean out its base and solder a red wire to the center and a black wire to inside the case. The wires should be about 1.5 inches long. Tin about a ¼ of the ends. This adapter allows the unit to be connected to the flashlight without having to split the knurled nut that holds the bulb.

39

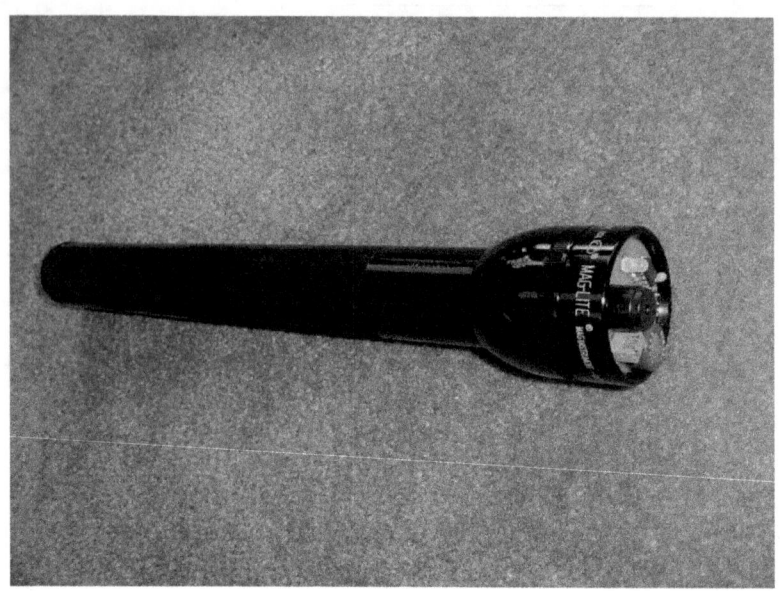

USING THE UNIT:

Remove the lens, reflector and the bulb from the flashlight. Place the bulb adapter into the bulb holder and screw on the knurled nut. Connect the wires to the board minding the polarity. Put the board into the lens holder and screw onto the flashlight. Prevent the board from turning by holding the components.

Turn on the flashlight. The led should light either red or green. The sound transducer may be squeaking. Let the unit warm up for about 2 minutes. Adjust the potentiometer until the led turns green. As the unit stabilizes the pot may need adjusting.

Open a rubbing alcohol bottle and place the unit near the top of the bottle. Any Booze bottle will work as well. The led should turn red and the speaker should change tone. The higher the tone the more part per million of vapor is present. When you place the unit in clean air the tone should fall and the led should switch back to green. The unit can also be used to check someone's breath to see if they have been drinking. Just have them blow towards the sensor from about 2 feet. The unit is, however, is qualitative and not quantitative. It does not mean the person is drunk. Need to check to see if the drink has alcohol in it? Just place the unit over the glass. Try it with a glass of wine.

Check out http://www.figarosensor.com/ for more specifications using gas sensors and their detection limits.

Happy sniffing.

CHAPTER SEVEN

DEMONSTRATION BOARD & DESIGNING A MICROCONTROLLER

Written August 13, 2009 for teaching twenty Moldova Bio-Engineers. Both my son and I taught this course.

PURPOSE:
1. Demonstrate the PIC16F72
   a. General programming
2. Demonstrate hall detector
   a. Bilateral
   b. Latch
   c. Unilateral
3. Demonstrate mounting of resistors
   a. Network
   b. Flat
   c. Standup
4. Soldering technique for surface mounts
5. Temperature measuring
6. Programming look up tables
7. Measurement of voltage
8. Op amp in non-inverting configuration
9. Programming for multiple segment displays.
10. Sound generation
11. **Using pull up resistors**
12. **How to program minutes and seconds**
13. **Programming interrupts**

SOLDERING THE BOARD:

The date is located on the bottom side of the board. The component side is on the top. The board should be rotated so that the seven segment displays are on the top and the dip switch is on the bottom.

Refer to the picture and schematic for parts locations located on the CD. *(See also SOLDERING: below.)*

    1.   Locate the three seven segment displays. The decimal points should be facing the microcontroller. Using a fine point soldering iron, solder each display to the board. Press the solder point to the board pads and the led connector and touch the solder to the pad. Each solder joint should take less than 5 seconds otherwise you can melt the plastic. *I strongly recommend using a magnifier when soldering surface mounts and inspecting the pads for cold solder joints.*

    2.   Solder the microcontroller to the board. Note: Pin one is the square pad. The notch of the microcontroller should be on this end. Solder the op-amp just to the left of the micro. The two notches should face each other. Snap in the Hall detector socket and solder.

    3.   Solder the red (reset switch) and black (start switch) on to the board. The red goes to left. Remove the film from the dip switch and solder it to the board. Note pin one to the square pad.

    4.   Solder the xtal, the two18 pF capacitors, .1 uF capacitor the 10uF and the speaker. Note the polarity of 10 uF cap and speaker. C2 and C3 are located next to the xtal.

    5.   Bend one of the wires of R1 (39K) 180 degrees and insert into its holes next to pins one and two of the op amp. This is known as a stand up resistor and is used to save space.

6.   Perform the same operation for three 10K resistors and place them in their holes next to the microcontroller pins 16-17-18.

7.   Solder the three PNP transistors into their respective holes. Their flats should face the right side of the board. The collector is pin 3. The board is marked with a C.

8.   Solder the remaining eight 10k resistors by bending them at right angles to the resistor. Resistor R5 is to the right of the reset switch parallel to the bottom edge of the board.

9.   Solder 470 ohm network resistor into its place. Note pin one.

10.   Solder the power connector. Solder the LM35 temperature sensor. The flat should face the top of the board

11.   Solder the 6 pin programming pins *inserting them from the bottom of the board.*

12.   *There will be six unsoldered holes. These are for future use.*

Inspect each solder joint. Check that inline socket notches face each other.

1.   Plug in a hall detector in the top socket. The bevels should be facing to the left.

2.   Set the dip switch so that the switches are all toward the top. (On, on, on, on.)

THE CIRCUITRY:

DISPLAY:

The three seven segment displays are placed in parallel with the exception of their anodes. An 8 resistor network (470 ohm each resistor) provides current protection to the segments and are connected to the micro. The three PNP transistors provide each seven segment display with 5 volts positive. The transistors have three current protection resistors (10K). The micro determines which segment should be lit and grounds the segments. The transistors provide power to each display. Each display is turned on for periods of mill-seconds. The human eye cannot detect frequencies above 60 cycles per second. The pulsing fools the brain to thinking that the all three displays are on at the same time.

SWITCHES & DIP SWITCH:

Each dip switch pulls its micro pins it is connected to ground. The pins are normally at Vcc via a pull up resistor.

The reset switch is connected to the master reset of the micro held to Vcc via a pull up resistor. When pushed it pulls the pin to ground. The same for the start switch.

HALL DETECTOR:

The hall detector has a socket to allow different hall detectors to be demonstrated. It also has a 10k pull up resistor to Vcc.

TEMPERATURE SENSOR:

The LM35 provides 10 mV per degree Centigrade. This voltage is amplified using the op amp which multiplies the voltage coming from the LM35 by 5. The op amp is configured as a non-inverting amplifier which uses the formula 1+ R2/R1 R1 = 10K R2 = 39K. The temperatures are displayed using lookup tables.

SPEAKER:

The speaker is connected to ground. Its positive pin goes to a 10uF capacitor which blocks the dc from getting to the speaker. The micro provides 5 volts + to the capacitor and generates ac to the speaker. The micro generates different frequencies.

XTAL:

The xtal used is 4.194304 MHz. This gives the proper timing signal for generating accurate signals for seconds.

RUNNING THE MICROCONTROLLER:

Plug in the power supply. The left side decimal should light indicating that power has been supplied.

Push the start switch, the demo board should count with each push.

Pushing the reset (red) button will set the microcontroller to its start point. You can then set the dip switches to another setting at any time. Once the dip switches are set, push the reset to reprogram. Then push the start button.

The following table is for the dip switch:

Dip switch	1	2	3	4	
1	ON	ON	ON	ON	Counter
2	OFF	ON	ON	ON	Timer
3	ON	OFF	ON	ON	Temperature in C
4	OFF	OFF	ON	ON	Temperature in F
5	ON	ON	OFF	ON	Temperature Register
6	OFF	ON	OFF	ON	Temperature Alarm
7	ON	OFF	OFF	ON	Hall detector
8	OFF	OFF	OFF	ON	Voltage
9	ON	ON	ON	OFF	Music Scale
10	OFF	ON	ON	OFF	French Siren
11	ON	OFF	ON	OFF	Siren
12	OFF	OFF	ON	OFF	New project
13	ON	ON	OFF	OFF	Future use
14	OFF	ON	OFF	OFF	Future use
15	ON	OFF	OFF	OFF	Future use
16	OFF	OFF	OFF	OFF	Future use

PROGRAM:

You can view the programming of the micro a couple of ways. I recommend that you load on your computer the MICROCHIP MPLAB from the disk I have provided. Unzip the MPLAB on the disk

and install. It is also can be obtained on the website
www.microchip.com . Please note that MPLAB updates about every
6 months

1.   Open MPLAB and select configure>select device and
      find the Pic16F74.
2.   Load the demo file from the disk.
3.   To turn on the line numbers click
      edit>properties>"ASM" File types and check "line
      numbers".

You can also view the code using notepad. I would also open
demo.lst to view the program addresses of the lookup tables.

The first thing you should notice is all the"; ". The "; "indicates that
compiler (the compiler converts the language into assembly code
for the final programming of the micro) should ignore the
information after it. It is just a way to write information for your
use and others of what is being done. Line 32 and 33 tells the
complier the type of PIC being used and what compiling programs
are needed to run it. Line 35 tells the complier how to configure the
micro.

_CP_OFF & _WDT_OFF & _BODEN_OFF & _PWRTE_ON & _XT_OSC
_CP_OFF   code protection off
_WDT_OFF   Watch timer dog off
_BODEN_OFF   Brown out off (if the voltage goes below a set value
               it turns off the micro)

_PWRTE_ON   Power up timer on (makes sure the power is
adequate on startup)
_XT_OSC   Xtal oscillator (depends on speed of crystal)

Note: Each microcontroller has its own set of configurations.

Lines 37-59 just gives information of what the inputs and outputs are and are only comments.

Lines 61-70 defines label names of ports and flag locations for the flag register.

Lines 72-88 names registers for various functions.

An ORG tells the complier to start the program at location i.e. ORG 00H start at zero.

The next command tells the program to jump to awaywego (a label)

ORG 4 is an interrupt location when the micro has an interrupt. The program for interrupts starts here.

Awaywego starts at line 143 and its following commands tell the micro what its ports are going to do, either input or output. Notice the ";", this tells what that particular line is doing.

Also you need to know, the majority of commands of loading registers have to go through the W register. You load the W register and then its contents are loaded into the register. The MOVLW B'00000001' means move the literal (binary 00000001 or a decimal 1) into the W register. The MOVWF PORTC means move the W register into PORT C.

START:

The beginning of the start program turns off the interrupts, and turns on one of the decimals to indicate that power is being supplied. The circuit then goes into a loop until the start switch is pushed. This switch is normally at Vcc via a 10K pull up resistor. When pushed it grounds RA5 and jumps to DIP_SWITCH.

The BTFSC SWITCH means "Bit test file (SWITCH) and skip if **not set**." The $-1 means go back one step.

The BTFSS SWITCH means "bit test file (SWITCH) and skip if **set**." This command eliminates switch bounce.

Once the micro is set up it will start its main program at start. I turn off the interrupt at numbers 173-175. I turn on the decimal and turn off all digits one and two to indicate power is being supplied. Label S1 waits until the start switch is pushed.

DIP SWITCH:

The dip switch goes to the RC inputs and is biased with 4 10 K pull up resistors to Vcc. When a switch is on it pulls one of the RC inputs to ground. The programming checks for the dip switch setting by subtracting RC from the W register. It then checks the STATUS register to see if the subtraction yielded a zero. If it did, the zero bit is a 1 and the next command is performed. If not, it is a 0 and the next command is skipped.

The BTFSC means "Bit test fr (file register) and skip if not set. The 3,2 means fr located at 03H (status register) and bit 2 (zero bit) . Review the "Instruction set summary" in the PIC16F72 instruction manual

Once the switch is pushed, the program goes to the label DIP_SWITCH. As you can see the labels give an indication of what the program is doing at that point.

Port C has the dip switch connected to it, but also has other inputs and outputs. To see what the dip switch is doing I used an "ANDWF" to only look at the dip switch.

COUNTER:

The first four commands clear several file registers to zero. The next four commands are checking to see if the start switch is pushed and released again eliminating switch bounce. With each push of the switch it advances the ones, when the ones rolls over to 10, it increases the tens and sets FLAG1. When the tens roll over 99 it increases the hundreds and sets FLAG2.

There are two types of jumping to another part of the program, the GOTO (to a label) and the CALL (to a label). The GOTO requires that after reaching the end of that part of the program it jumps to, another GOTO (to another label) is required.

The CALL (to a label) has the advantage that many parts of a program can call on the area and it can be used over and over. There is a RETURN at the end of the CALL and the program counter goes back the next step from where the CALL originated.

The call display jumps (line 262) to the label display. Now its program runs. Note there is another call for PATTERN (line 312) which is a lookup table. The ones are place into the W register and when the jump takes place the W is added to the PCL and the pattern for PORTB is returned for display. The digit is then turned on. The tens and hundreds result is performed in the same way. Remember that the micro is running at 4 Hz and the eye cannot detect the change of each digit being turned on and off. This is called multiplexing.

TIMER:

The timer program allows the timer interrupt to be turned on. A decimal 16 is loaded into LOOP for use in generating one second signals from an internal micro timer. Several registers are set to zero and the flags are all set so that all the numbers are on. Timer1 program is a loop which causes the seconds and minutes to display. When an interrupt happens, once its program runs, it always returns to the place in interrupted.

This microcontroller has 8 interrupts. When an interrupt happens it stops the program and jumps to memory position 04H. At this location there is a GOTO VECTOR. In the VECTOR area the first six and the last six commands are housekeeping to help the micro go back to where it was interrupted from. There is an instruction to stop the interrupts.

The crystal was specially picked at its odd frequency of 4.194304MHZ. The microcontroller counts every 4 cycles so the frequency is now1,048,576. One of the interrupts in this micro is called the TMR0. It counts down from 256. In the OPTION register, you can program the TMR0 so that it does this 16 times before it causes and interrupt.

1,048,576 is now divided by 256 = 4,096 divided again by 16 = 256. There is another counter added to the program called COUNTER and this divides again by 256 which will produce a one second count.

The following instructions just count seconds, tens of seconds and when it reaches 60 seconds it counts minutes.

After the housekeeping there is a RETFIE which returns the program to where it was interrupted and turns on the interrupt.

By adding other counters you can count micro seconds, milli-seconds, hours, day's months and years.

TEMPERATURE IN C:

The temperature C program, the A/D converter is turned on using the ADCON0 SFR register. It is started by setting the go done bit.

This bit is watched until the A/D conversions over and the go_done bit goes to 0. The result of the voltage of the LM35 is proportional to temperature. This voltage is multiplied by 5 by the op amp and stored in the SFR register ADRES. A call is made to the lookup table and the result in centigrade is brought back for display. The lookup table address is located at the program address of 04F0H (line 898). The lookup table was set high in the program counter to provide for the maximum amount of program commands. The result placed in the W register is in hexadecimal and needs to be converted to decimal. This is done by decreasing the count while counting in ones, tens, and hundreds how many decreases it takes to reach zero. This result is then displayed.

TEMPERATURE IN F:

The temperature program for F is identical to C however a different lookup table is used. The address of the lookup table is located at 06F2H (line 1423).

TEMPERATURE REGISTER:

The temperature register displays the result of the A/D converter from the ADRES register. It does not need a look up table.

TEMPERATURE ALARM:

The temperature alarm uses the ADRES register and determines if the register is above or below given numbers. The leds are programmed to display an L G or H for low, good or high. Lines 561 to 568 determine the range. You can use an ice cube or a heat source to view the alarm.

HALL DETECTORS:

The hall program checks for either a transducer producing Vcc or Vss (high or low 1 or 0). The hall detectors are connected to Vcc and Vss. They act as a switch and are often used for the control of positioning of belts, motors, doors etc. Most require a pull up resistor and are connected directly to a port. The BTFSC HALL_DET command (Bit Test File and Skip if clear) is the heart of the program. The rest of the line codes are just for displaying ON or OFF.

There are three hall detectors for demonstration, unilateral, bilateral and latching.

The beveled edges should face the left side

VOLTAGE:

The voltage program requires that a voltage is placed between ground and pin 3 (RA1) of the microcontroller. There are two solder points for this application. The lower one is ground (just below the resistor) and the other goes to pin 3. Do not exceed Vcc. A pot can be used for demonstration by placing Vcc on pin one Vss on pin three and placing pin two (variable voltage) to the micro pin 3. By varying the pot you will be able to see the voltage divider in action.

The program starts off by telling the micro to use RA1 as the A/D input. It is identical to the temperature centigrade program but uses a different lookup table. The lookup table converts the ADRES register in to volts. The lookup table is located at 05F1H.

MUSIC SCALE:
The music scale program demonstrates the sound capability of the micro. By pressing the start switch the C3 scale will be played and will advance each push of the button. This is done by turning on and off the speaker at different calculated rates.

Counting registers in each NOTE label determines the length of time the speaker is turned on and then turned off. Notice how the registers change. This program produces a square wave which is full of harmonics. You can view these square waves by using an oscilloscope connected to ground and pin 15 (RC4)

FRENCH AMBULANCE ALARM:
This program uses two tones alternating between note C and note F.

SIREN:
The siren program loads registers with predetermined counts. The results of these registers are transferred to other registers which are decremented for a predetermined times. The tone and the speeds can be changed by changing the numbers.

NOW IT'S YOUR TURN:
There are five more dip switch settings for writing your own programs. You can experiment using different switch, modes, timer

modes, temperature alarms, voltage alarms the list is endless. You may want to erase some of the programs for more space as the program memory ends at 02B7H and the lookup tables start at 4F0H leaving 269 programming spaces. I would rename the file and save it as an .ASM file.

The first thing to do is perform as save with a new name. You will need to add an .asm for it to work.

Try something simple to keep your frustrations down. Write your code after NEW_PROJECT. Get rid of the GOTO $ and write your program.

Suggestion; Try writing a program that uses the hall detector and plays a note when the hall detector is turned on. This can be done using 7 commands. Don't forget to compile. Review the Hall program and the scale program for help. To program the chip do the following:

Use the PICkit 2 and place the programmer arrow to the programming pin closest to the top of the board.

DESIGNING YOUR OWN PROJECT

PROJECT:
1.   Analyze your project to determine the number of inputs and outputs you need.
2.   Are you going to measure a voltage with a sensor thus requiring an A/D converter (analog to digital)?
3.   Do you need to amplify the sensor and is the voltage +- from ground? This will determine your voltage supply i.e. op amp requiring a dual supply.
4.   What outputs are you going to need? Relay triac, sound, leds, motor, etc.
5.   Do you need a voltage output thus requiring a D/A (digital to analog) output?
6.   What power are you going to use? Battery, 220, 110, consider using a switch mode plug in battery eliminator.

MICROCONTROLLER:

1. Go to www.microchip.com and use their product selector to determine what cheapest and lowest pin count you can use (Help me select a part). I like to use the F series (the F stand for Flash i.e. 16F74) if available as they can be reprogrammed over and over and can be used in the circuit without having to buy a onetime programmable. The C series (i.e. 16C71) requires that you have a UV eraser and you have to buy the windowed version to reprogram over and over. After you have the proper program you buy the one time programmable version and use this as it is cheaper.

2. A/D comes in 8 bit, 10 bit and 12 bit. Keep in mind if you have a lookup table they can get big as the bits go up.

3. Many of the micros have internal pull up resistors. When using switches going to an input, you need a pull up resistor to Vcc to keep the gate high. When the switch is pushed the gate goes to ground. Internal pull up resistors eliminated an extra resistor on the board. Many photo detectors, hall detectors can go directly to the gates with internal pull up resistors.

POWER SUPPLY:

1. The easiest power supply is battery but keep in mind that if you have leds on the battery supply can get big. Consider putting the microcontroller to sleep during part the program as the power consumed is about 5 micro amps. Leds pull about 10 milliamps and but can be flashed to save power.

2. The majority of micros can run at 4.5 volts thus only requiring three 1.5 volt alkaline batteries. You can use 6 if you have an Op amp needing plus minus five volts (4.5 actually).

3. You can use a 9 volt battery but its voltage will have to be dropped down to 5 volts. If the project is only on for a few minutes, I use a L7805 (100 ma). Keep in mind though; the other 4 volts are reduced by heat dissipation. Most people forget that if the project is

using 20 ma @ 5 volts it requires .1 watts for the circuit, however the battery has to provide .18 watts as .08 watts are given up in heat for the 4 volt drop.

4. You can use a 9 volt battery with a switch mode chip and get better efficiency. I use a LT1121 and all you need is a capacitor. You can also generate negative voltage take a look at LT1175.

5. I recently have gone to a plug in switch mode power supply. Take a look at DigiKey T946-P5P-ND 5 volts at 1200Ma with different country adapters. Both 110 and 220. It saves board space and costs $14.00

BOARD LAYOUT:

I want to warn you that you will make mistakes on your board layout, often reversing pins, having open traces or crossed traces. This can be frustrating and I still do it after laying out several hundreds of boards. When laying out boards, the bottom side is viewed and if having x-ray eyes via the top side. This will be reversed automatically by the board making company.

I use the free software by Express PCB
http://www.expresspcb.com/

They provide three 3.8" x 2.5" for $60.00 USD. They also have other sizes. Once the board is designed you just click on Layout and send the boards over the internet and they will mail them back to you. They will also make any size board for you for different prices.

If you have very small boards you can use the copy function and I have placed as many as 6 boards per single board thus giving you 18 boards for $60.00.

Once I have determined the parts I'm going to use, I copy all the PDF files to a folder named for the project. In this folder I keep the board layout, description and the parts list with pricing and part numbers.

I then print out the foot print of the parts. This will help to eliminate part reversal of pins. Don't worry about the traces going in order to the micro. I.e. segment "a" going to RB0 a segment "b" going to RB1. Rout the traces for your convince and change the pins when programming the micro. Also if the routing may be better if you turn a chip so that the notch is to the right instead of to the left.

1. Keep in mind you will need mounting holes for mounting the circuit board to a box or plate.
2. You can also use parts for mounting the boards, transistors, switches etc. I.e. when using optical triacs, I place them on the edge of a board parallel to the board and heat sink them to a chassis. This saves board space.
3. For tight boards you can stand the resistors up on end instead of using them flat. If using a lot of the same resistors i.e. 7 segment leds consider using SIP packages. They come in different pin counts and configurations.
4. I always use IC sockets for the prototype board as you will find when programming the microcontroller you will be putting it on and off the board many times. Get a chip puller to prevent bending the pins when removing the chip.
5. If you have the parts available, I often place them on sheet of paper the size of the board I think I'm going to use and move them around.
6. If I have extra gates, I often place an extra led with a resistor on board so when I'm programming I can see where I am if I have a glitch in the program by turning on the led.
7. Check the part specifications for the size of the pad and the drill hole.

Traces:

The following table is for determining the size of the trace to the amperage it has to conduct:

Size inches	2.0 ounce trace
.005	.7 amps
.010	1.4 amps
.020	2.2 amps

.030	4.0 amps
.055	5.0 amps
.150	I.0 amps
.370	20 amps

The following table is for determining the size of trace spacing for voltages:

Volts	Spacing in inches
	.015
	.035
	.050
	.1
500+	.0002 in/volt

Once you think you have the board laid out, print out both sides of the board.

Using different colored markers, trace out the Vss and Vcc traces and check to see if they are going to the right pins.

On your parts list, make a list of what the pins are connected to.

You will need this when you program i.e.

RA0	Analog in
RB0	Start switch (low to start)
RB1	Door switch (low closed)
RB2	Hall detector (high when magnet is detected)

Have the board processed.

SOLDERING:

1. I use an 800 F degree .031 long conical tips Weller soldering iron. .02 or .031 solder.
2. Make sure you heat the pad and the part before adding the solder. Touch the solder to the part and **NOT** the tip of the soldering iron to melt the solder. Slide the solder tip up the part pin. This will pull the solder and insure you don't have a cold solder joint. The solder joint should shine and not be dull.
3. Most of the time when removing a part I use Chem Wik .030" for small parts and work up to the larger sizes of Chem Wik for bigger parts. Place the solder Wik on the pad

to be de-soldered and heat the solder Wik. Once you see the solder melt, pull the solder Wik to wick up the solder.

4. If the Chem Wik does not pick up or leaves the hole full of solder, I add solder to the hole and try again.
5. I also sometimes use a solder pump to clean out holes.
6. Use an ohm meter and check out the Vss and Vcc are going to the correct pins on the board.
7. The first time I try a new board, I place a .5 amp fuse and an amp meter in series with the power supply. If the chips pins have been reversed on the board, the fuse will blow and you won't lose the part. The meter tells me how much power is being consumed.

PROGRAMMING THE MICROCONTROLLER:

Now for the fun part and the most frustrating!! There will be mistakes, I have never written a program that worked the first time without errors.

Down load the latest version of MPLAB from www.microchip.com and install.

1. Get the data book on the microcontroller.
2. I prefer to program without using their wizard as I never got it to work correctly.
3. Open MPLAB and select configure>select device and put in the number of the device.
4. Import sample program I have put on the disk. This will get you started.
5. Fill in data in the sample program. *(View the demo program for examples.)*

    a. Name the program

    b. Fill in date, version, and xtal type, written by.

    c. Look in the data book and get where the general file address start, the interrupt address, and the last memory location.

    d. Add as many comments you need for future use

e.   Put in the microcontroller number i.e. p=12c508

f.   Put in the microcontroller number, i.e. <p12c508.inc>

g.   Perform a search on your computer for the ?????.inc file of your microcontroller and print out. (i.e. search for 12c508.inc) Look at the Configuration Bits. *(View the demo program for examples.)* Each configuration bit is started with a _ and a & is place between them. The order does not make a different.

h.   It's a good idea to list all your ports so that you don't miss any

i.   Define your ports. *(View the demo program for examples.)*

j.   Put in your register file names *(View the demo program for examples.)* The address depends on your microcontroller. Look at its data sheet under general file registers.

k.   I always start at 0H and jump to the starting address. This leaves room in case you want to have an interrupt.

l.   Write your code.

m.   You always must have an end at the end of the program.

Make sure you save your code as an .asm file

COMPILING THE CODE:

Make sure you have put in the microcontroller in the "Configure"

1.   Click <Configure<Select device and find the device.

2.   Click <Project<Quick Build .asm

3.   The Debug screen will list your errors, comments and warnings.

4.   Now all you have to do is correct the errors. (At first there will be a lot!)

5.   Once you get a BUILD SUCCEEDED you can burn the microcontroller. You will need a programmer. *(Refer to instruction manual that comes with the programmer)*

Once compiled with a Build Succeeded you will end up with 5 Files generated

*.asm	The assembly code
*.cod	Compiled source code file
*.lst	The listing file used by the debugger
*.hex	The hex file use when burning a chip
Error log	File containing errors or warning

DEBUGGING:

Don't be surprised after you burn your microcontroller it does not work. Try small programs first and then build. The MPLAB comes with a debugger called MPLAB SIM .

You can start the debugger by clicking <Debugger<Select Tool<mplab sim. Review the help file included.

A couple of tips:

1.   If using the stopwatch, make sure you put the correct crystal frequency.

       a.   Make sure the debugger is on.
       b.   Click <debugger<Settings and put in the frequency.

2.   Use the watch (under view) to change file numbers. Keep in mind that they are in hexadecimal.

3.   To view ports and the W register use the Special Function Registers (under view) .

4.   To change port settings, you have to use the stimulus program. (See help file)

5.   To set a break point, click the right mouse button and click set break point. Break points will not be set on a label if there is no code beside it.

PROGRAM MEMORY AND REGISTERS:

Each micro has its own program memory for writing code. Program memory has nothing to do with registers and sometime cause confusion with addresses. The program memory starts at 00H which is also the reset location meaning if the micro crashes the program will start here. 04H is the interrupt vector. The micro will jump to this location when there is an interrupt. The pic16F72 has 8 interrupts capability. Other Pics have less or more. Any time there is an interrupt, the program jumps to vector 04H. You can either tell the processor to jump somewhere else or start writing the interrupt code there. Notice there are four open places to write code before 04h. I normally put the jump to awaywego (just a label I thought up) at 00H and write the vector code starting at 04H.

The Pic16F72 program memory ends at 0800h which allows for 12,800 lines of program. Other Pics have can have more or less memory.

The Pic16F72 have two types of registers. The special function registers (SFR) most of them can be written to but there are some that cannot. The other type of register is known as General Registers. These registers start at a specific location. This location varies with each type of Pic and you have to look it up. Not looking it up can place your program in a non running condition or worse yet unpredictable chaos. The Pic16F72 general register starts at location 20H. See line 72. Registers *have nothing to do with program memory!!*

There are 128 bytes of general registers you can use. You can name these almost anything you want. If you are going to use two names use a "_" to join them. I.e. RONS_COUNTER.

If you are going to write to a SFR i.e. PORTA, make sure you are in the right bank. This can be done by using the command BANKSEL PORTA (bank select port a). The debug when compiling will flag these areas but will not correct them.

LOOKUP TABLES:

Lookup tables can be very confusing especially if you have 255 (FF) entries. You must always keep in mind that registers can only count to 255 (8 Bits), if you are going to count higher you must use another register. The first command in a lookup table is to add the offset (now in W) to the PCL (program counter low). If the beginning of the lookup table starts at 300H the first lookup is at 301H. If the offset is FF, this will cause the PCL to roll over to 00, but the PCH (program counter high will not increase. This will cause the program to jump to some unknown destination which is not good. The easiest way to prevent this to use the following which will allow you to put a lookup table any place in the memory code.

Let's put a lookup table at a memory location of 4CCH by using an ORG 4CCH at the beginning of the lookup table.

The offset will be FF.

movlw	CCH	;This is the lower bytes of the ;memory location
addwf	offset,0 ;	The offset in the number you ;want to add to the PCL
movlw	.4	;This is the high byte
btfsc	status	;this checks to see if the PCL ;rolled over
addlw	1	;add one if it did to the ;PCLATH
movwf	PCLATH	;PCLATH addresses the PCH ;(you can't program PCH)
movf	offset,0	;put the offset in to the W ;register

63

```
call                TABLE

org     4CCH                    ;Sets the table at a location
```

TABLE

```
        movwf       PCL,1           ;add W to the PCL
        retlw       'A'             ;return the ASCII char A
        retlw       'B'             ;return the ASCII char B
        retlw       .10             ;return the decimal 10
```

If you are going to use several lookup tables one after another, after you assemble the code, view the *.lst (listing file) and find what the memory code has been assigned to the lookup code. Go back and put in the PCL and PCH numbers in the area you are going to call the lookup table

NUMBERS:

There are several ways of loading different types of numbers:

Binary ports	B'01100100'	often used when setting up
Octal	O'144'	rarely used
Hexadecimal	H'64'	
Decimal	D'100'	
Or	.100	Easiest to use
ASCII	A'd"	

Note: all the above = a decimal 100.

FLAGS:

Flags are used show that an occasion has occurred. In the demo project flags were used to indicate that that digit should not be displayed. It loaded a 0AH which jumps to no segments lit in the pattern. I assigned and address for a FLAG register. Since you have eight bits per register, you can have 8 flags.

Happy programming!!

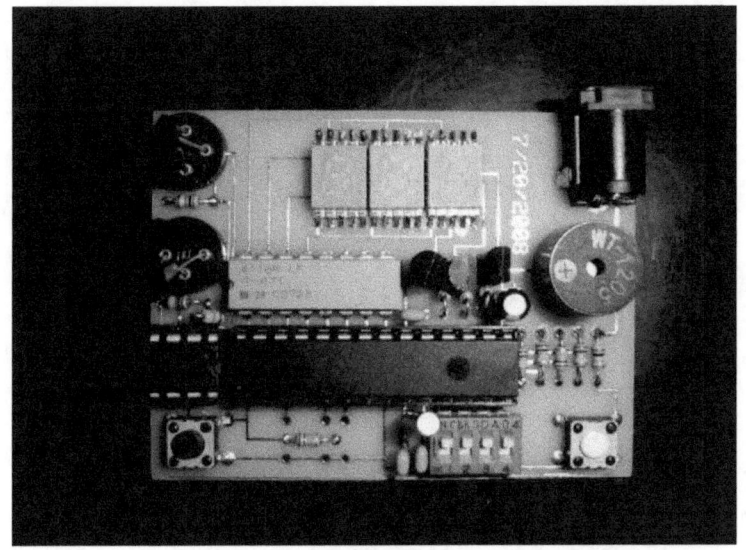

DEMO BOARD

CHAPTER EIGHT

DETECT THE EARTH'S MAGNETIC FIELD CHANGES WITH A TORSION MAGNETOMETER

William Baird's September was a good article and used a hall detector as a magnetometer. His device was designed for measuring fields of 10 gauss or 10^{-3} milli-Teslas and was good for magnets and components. To determine the effect of sun spots you need to get down to 4 nano-Gauss or 4 nano-Teslas (10^{-9} Teslas) a magnitude of 1,000,000 times more sensitive.

With the sun spots being abundant this year (2013), the earth's magnetic field changes rapidly. This can be detected using a set of magnets and mirrors. Many articles are on the internet on how to build one of these. Most use a laser that hits the mirror which its reflective spot moves on the wall. The principle is an old one and the ancient galvanometers used a lamp, magnet and coil to measure current. I have adapted several ideas so that this one can be a small device and a standalone unit with an LCD displaying variations. A digital to analog converter (DAC) is also included on the board allowing you to plug into a chart recorder or better yet a computerized data acquisition logger. If you want to use it in the field you can use a field data logger voltage recorder. The logger will chart the data in graph form.

66

This device is so sensitive it can detect when your neighbor drives his car into the garage. It can be built for about $40.00.

PRINCIPLE:

The principle is simple. A compass points to magnetic north. It is difficult to see that the needle changes as the magnetic field changes. These changes are minute. If a magnet is suspended on a small filament with a mirror attached to it, it will also point north. When a beam of light is shown on the mirror the mirror will reflect the light. Small variations in the angle of the mirror will be amplified as the surface distance varies from the mirror. Many of the articles use the wall and ruler to measure the changes. Instead of a wall, a linear array detector with 128 photodiodes detects the change of movement of the beam. The microprocessor detects which diode is being hit. The beam is calibrated to hit diodes 64 or 65 which are arbitrarily set at +1 & -1. If the beam strikes the diodes to the left, it will show a positive displacement up to 64 counts. If the beam strikes the diode to the right, it will show a negative displace up to 64 counts. The DAC converts the diode number from 0 to 128 into a dc voltage. 5 volts being diode 128. The chart software can center the 2.5 to a zero reference so that you can see the change both positive and negative.

The construction for this Magnetometer will detect magnetic variations of 5 degrees swing total. Each pixel detects .04 degree of change. If you live in northern Canada, some magnetic storms can produce greater than 5 degrees deflection. The 128 pixels are an array measuring 1.6 cm and sensor to the mirror was set for 9 cm. For those who are interested, the formula for calculating the angle displacement is:

$$Angle\ in\ degrees =$$
$$\frac{1}{2} * 57.307\ x\ \frac{deflection\ in\ centimeters}{distance\ in\ centimeter\ from\ mirror\ to\ sensor}$$

Note: 1/2 is due to mirror doubling the angle.

LINEAR PHOTO DIODES:

The nemesis to this device is stray light. Mark pin 1 on the back of the chip for a reference. Cut a 1.5" x 1/8" strip of masking tape and place over the diode area. Coat the front and sides with black paint or black nail polish. Make sure the corners are covered. Use two coats if necessary. Once dry remove the tape.

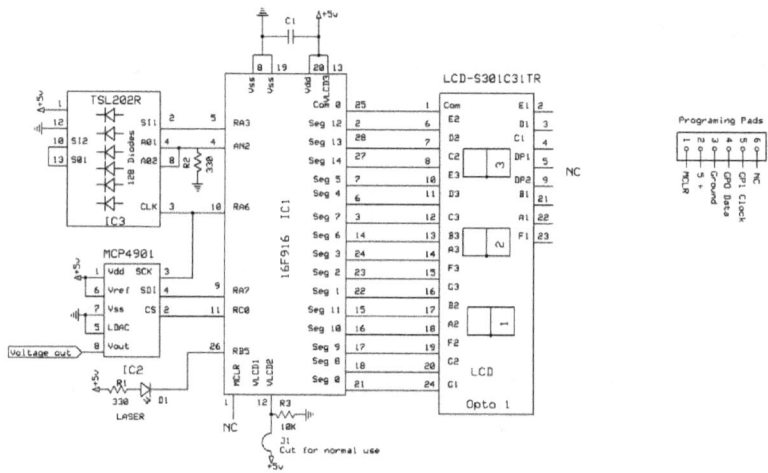

BOARD CONSTRUCTION:

If you want to make your own boards, the Express PCB files along with parts source, microprocessor code and "Hints and Tips" are located on the CD.

Mount and solder IC1, IC2 and the LCD display with pin 1 going to the square holes. Solder R1, R2, R3, C1 and the switch. Solder the 14 pin vertical socket facing the edge of the board. Plug in the linear array with pin one to the lower left. Pass the battery wires through the strain hole and Solder to the pads labeled + and -, the red wire goes to the positive. Solder two pins if you are going to use a chart recorder, data acquisition or a data logger. These can be used with wire wrap or a female jack. Cut the small trace with a razor for normal use. Trim the socket pins and the LCD pins so they are flush with the board. The laser will be mounted later.

SOFTWARE & ELECTRONICS:

The laser power has been decreased by R2 and you will barely be able to see its line. This was done as the Taos linear array is quite sensitive and stray light reflecting from the diodes will cause the chip to saturate. The clock for the linear array is running as fast as the microprocessor can handle for short integration times. The laser is turned on and off to consume power.

An algorithm is used to see which pixel has the highest voltage. The pixel number is stored in memory. Once all 128 pixels have been measured, a lookup table is used to determine the pixel number from the center of the display and is constantly being displayed on the LCD. The Pic16F916 was designed for driving LCD displays. The DAC takes the pixel number and converts it to an analog voltage.

There are two modes which the unit can run. The first allows sampling at a rate of 166 samples per second. This is ideal if looking for sudden changes in magnetic fields. The second is for field use and saves power. It samples once at one minute intervals and can be used with a data logger.

MODIFICATION OF LASER

I tried a number of laser diodes including IR. I came across a Harbor Freight saw laser marker for $4.99 (Item #93242). It is ideal as it draws a laser line and not just a dot. I don't know how they can sell it for this price, as the batteries that are included are worth that much.

Remove the end plastic by using a screw driver to pry off. It is not glued just a pressure fit. Using a Moto tool with a cutoff tool, cut the plastic about 1/8" behind the brass. There is brass behind this so you don't have to worry about cutting into the circuit board. Cut until you hit the brass.

File, sand or use a lathe to turn the brass end piece off the proximal end so that it is even with the laser tube. Remove the washer and O ring. Remove the three screws and take off the collar. Solder a red wire to the brass. Cut the spring and solder a black wire to the solder that holds the spring.

Insulate the brass by using shrink tubing allowing the wires to exit the distal end. Check to make sure the diode is working by using 3 volts.

PVC ASSEMBLY:

All parts can just be pushed together and they don't need to be glued. The chassis for the magnetometer is made of 1" PVC pipe with one cross, one tee, one cap and 4 plugs. Cut 2ea 1.75" pieces of 1" pipe and deburr.

Using a lathe or sandpaper remove about .002" from 1" of one side of the 1.75" pieces. The end cap should slide on easily. Push the tight end of this pipe into the arm of the tee and slide on the cap. Take the other 1.75" and push it into the other arm of the tee. Press a plug all the way into the bottom of the tee.

Take two plugs and using a 1/16" drill, drill a hole in the center of each. These will hold the nylon filament. Press these all the way into cross opposite each other.

Now join the cross and the tee using the non-modified 1.5" pipe. The cross should be vertical. Press the two plugs with the 1/16" holes opposite each other into the cross. The plug in the tee and the bottom plug of the cross should be the same height as this is the stand. Mask off the inside of the empty cross hole. Using flat black paint, spray the inside of the cross and tee to help prevent reflections. When dry remove the masking tape.

Take a plug and using a lathe or sandpaper remove .002". It should slide in the cross with little force. Drill and tap it with a 1/4-20 hole in the center. Make the damper out of a penny minted before 1980, (better yet an aluminum slug) by gluing the head of 2" 1/4-32 flat head nylon screw. A hot glue gun or super glue will work. Make sure it is parallel and in the center of the penny. Place this attenuator inside of the plug. The penny goes into the hole even with end of the plug Add the 1/4 - 20 nut to the outside of the plug. This will act as jam nut.

The mirror that is used should be a surface mirror as refraction will occur when passing through the front glass of the mirror. Your local craft store carries .75" round mirrors which are ideal for this project. You can get 10 pieces for $.99. In a well ventilate area; take a mirror and using lacquer thinner remove the backing using a Q tip taking care not to scratch the surface. This may take several tries as the aluminum coating scratches easily.

Using a compass, take the pair of magnets that are snapped together and determine the South Pole. It should be attracted to the North end of the needle compass. Superglue the glass side of the prepared mirror to the south side of the magnet making sure it is centered. The mirror surface should be on the outside. Now put a drop of superglue on the other magnet and glue the glass side of a non-modified mirror to it. The coating of the mirror should be facing outward. Let dry. This mirror is for balance.

Take 1.5' of nylon twine and separate its strands until you get one small strand. Using masking tape, tape both ends so you know where they are. Slide the mirrors apart. Place the nylon strand in the center of the magnet. Hold the mirror with one finger and take the other mirror and let it snap together with the other magnet impinging the strand. Make sure it is centered!

I used two 6" pieces of different colored wire wrap to avoid confusion. Tie a knot on the end of the top wire to prevent it from falling through the hole. Place 2 ea 7 gm split shot (about 7/16" diameter) on the bottom wire. Thread each into the holes in the plugs and pull them out the remaining 1" hole of the cross. Put about 4 wraps of the strand around the wire coming from the top plug and add a drop of superglue. Gently pull the mirrors into the cross and center them in the hole. Use masking tape to hold the top strand. Now repeat the procedure for the bottom strand. Hold the cross with the monofilament line vertical allowing the weight to tension the strand. Double check to see if the mirrors are centered.   Use a toothpick to lock the top strand. Once the mirrors and the sinkers have stopped spinning, use a toothpick to lock the bottom strand.   Cut the tooth picks flush. The magnet with the mirror should be pointing to the south.

CIRCUIT BOARD MOUNTING:

Mount the board upside down in a vice and pick a spot on the wall about 1' from the vice. Using a string, tape it to the center end of the board (LCD end).Pull it down the center of the board and to the mark on the wall. Align the board so that the cord is in the center. Remove the string.

 Using 2 AA or AAA batteries in a holder as a power supply, connect the laser wires to batteries. Using a hot glue gun put a 1/2" x 1/8" bead of glue in the center of the board where the socket is. The laser line should be perpendicular to the board. Push the laser into the glue with its end even with the end of the linear chip. Adjust the beam so it is vertical and intercepts the spot on the wall. Let cool.   Disconnect the laser and solder its wires into the R and B pads.

Take the cap and using a mill or Moto Tool, cut a horizontal slot in the end. See the CD for the template.   The LCD should slide through this slot. Once you know everything is working, you can fix the bottom of the board to the plug using hot glue from the inside. Slide the board and end cap into the tee.

FINAL USE:

Put the damper into its hole. Using a compass, align the tube north and south with the LCD on the south end. The unit consumes about 30 ma. Use 3 D cells for power or a 5 volt battery eliminator if running a chart recorder. Turn on the power. Adjust the tube right or left until you get a near 1 or + 1 reading. Make sure that you give time for the mirror to settle down. This may take several minutes up to an hour. If you still have oscillations screw in the damper until they are damped . Use the jam nut to prevent the damper from changing. If you want the unit for field use, solder the two pads J1.

Magnetogram:

I have several chart recorders that I use throughout the years. These use reams of expensive paper. I have replaced them with my old computers and data acquisition units and software. Everyone ends up every few years replacing their old computer. They make ideal chart recorders. Dataq (www.dataq.com) makes a data acquisition module which sells for under $30.00 and measures 4 analog signals along with two digital inputs. This is a must for any scientist's laboratory. I have been using several of their products for 20 years or more. You just connect the module to your computer and it charts your data. With their free software you can compress a month of data into one page for printing.

The DAC produces 0-5 volts. Connect the data acquisition unit by using a jack or wire wrap on the ground and analog pins. Plug it into your computer and call up the software. There are a plethora of commands just refer to the software manual. It will monitor on the screen a live display of the voltages. It will sample up to 240 Hz giving you a magnetogram. I have used these types of modules for seismographs recording, recording a month of data at a time. You can save the data and call it back looking at individual periods of one minute.

For field use and stand alone, Dataq makes voltage monitor that plugs into a USB port and charts the results. The EL-USB-3 USB Data Logger sells for $75.00 (www.dataq.com click on USB data loggers). It will sample from one second for 9 hours or every minute for 22 days or for other applications sample 12 hours for > 2 years.

Start the EL-USB-3 USB Data software and set its parameters. You can use an immediate start or a delay start. It even has alarm led's. Connect it to the red and white wires noting the polarity. You can remove it at anytime and download its data using the USB port and it will chart your data for you.

Some people just have magnetic personalities not due to sun spots.

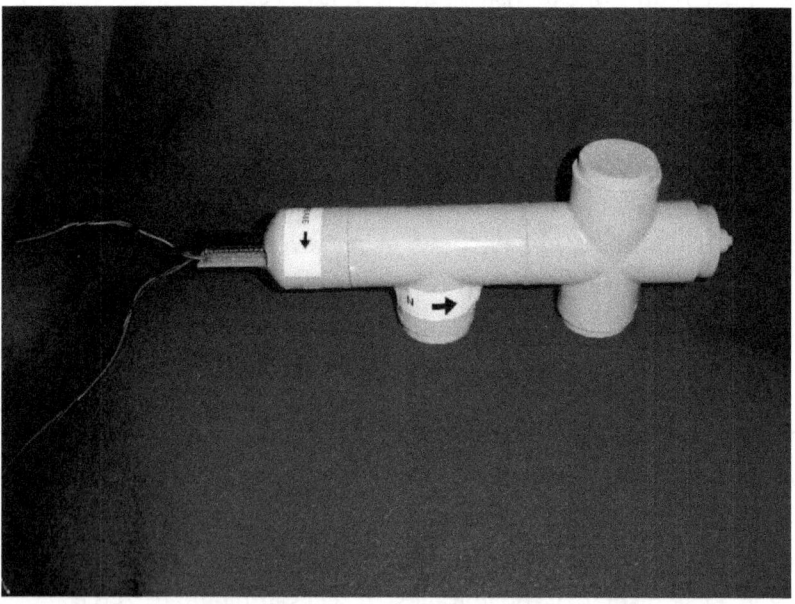

See CD chapter 8 for parts and source list and more pictures.

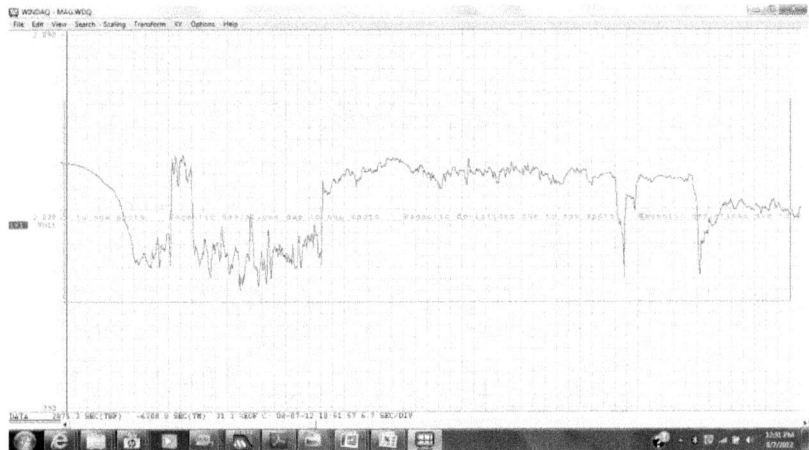

CHAPTER NINE

THREE AXIS HOCKEY PUCK ACCELEROMETER DATA LOGGER

By Ron Newton
May 12, 2012

Accelerometers can be used in many applications. This one was designed as an adjunct to the "Poor Man's Seismometer" published in the Nuts & Volts May 2012 addition. For those who are studying vulcanology, accelerometers can be used for determining which direction the magma is moving using the Z-axis. The sensor used measures in three dimensions using a "Freescale Semiconductor" ±2g/±4g±8g three axis low g digital output detector. It can be built for less than $20.00 plus board.

 I have designed this project so that it can be programmed using the SD card as a "Seismograph", a "Continuous Monitor" or a "Shipping Monitor". The data collected on the SD card can be downloaded into a computer and displayed using Dataq Web browser, which is free software from Dataq. All three dimensions are displayed at the same time.

Ever wonder how hard the post office or UPS throws your package marked fragile? This device will show you. Its tripping point can be set from 0.1 g to 8g on all three axes. It records normally for 5 seconds and then goes to sleep. The recording time can be changed in the Excel spread sheet. The advantage over static g monitors is that this one dates and times the event.

For a "Continuous Monitor" all three axes are monitored on any of the setting of 2g, 4g or 8g. This is also set using the Excel spread sheet. It will continue monitoring until the program button is pressed. It will record up to 256 events.

The continuous monitoring, has many applications e.g., racing cars, model airplanes, rocketry etc. The SD card size will determine how long it will record. It records 80 samples per second. It is a great device for prototyping shipping containers. How about a physics project of building a container that contains an egg and dropping it from a height of 20 feet without breaking it? This device will tell you the forces involved.

In the Seismograph mode, once an earthquake above a .05g is detected on the X or Y axis, it records all three axes on to the SD chip for one minute using the 2g sensitive mode. As far as using it as a seismograph, it will not be as sensitive as the "Poor Man's Seismometer" but will work for larger earthquakes showing the direction of the waves and their force.

There are two main types of scales for measuring earthquakes, The Richter and the Mercalli. The Richter is logarithmic developed in 1935. Each number represents 10 to that power. An earthquake of three is about what most people can feel. 10^3 = 1000. The San Francisco quake in 1906 is estimated to be an eight. 10^8 = 100,000,000. This is 100,000 times more powerful than a three.

The Mercalli scale is an arbitrary scale based upon what people feel. It was revised in 1902 and is a 12-point scale. The San Francisco quake equates to an X or XI on the Mercalli scale.

Gravity is the force pulling things down and accelerates objects at $9.8 m/s^2$. For something to leap in the air with an earthquake, requires a force wave traveling above 9.8m/s^2 (1g) which is equivalent of an 8.1 on the Richter or an 11 on the Mercalli scale.

The sensor used in this article can measure down to .016g or .0.16m/s$^{2.}$. This would be an IV on the Mercalli scale or about a 4 on the Richter scale. The accelerometer is more akin to the Mercalli scale than the Richter scale. What the accelerometer does over the seismograph is to give you a three dimensional view of the g forces involved. This is an engineer's dream tool for studying structural engineering.

The sensor measures 3x5 mm. Its cost is minimal, less than $2.00 ea. They are quite difficult to solder and there is no use in ruining the larger board, which holds the microprocessor and SD card so I planned to mount them on 8-dip header. If you wonder why the chips are so inexpensive they are mass produced for hand held computers and detect if they are shaken, jarred, rotated or tapped just by using g forces. They use two capacitors with a plate in between and measure the change of capacitance between them. The MMA7455L has an analog to 8 bit and 10 bit converter and transmits the data using I2C protocol. By using a twos complement it gives positive or negative g forces. The first seven bits give 2^7 = 128 bits of information. The eight bit changes the sign and the following data bits are another 128 bits of information. The unit has a built in clock which its time is programmed with the SD card. There is a power switch and push button switch that programs the unit. A bi-colored led acts as an indicator for programming. When the g force exceeds a predetermined amount, it starts the recording, and indicates by a flashing led that data has been collected. 2 AAA batteries will provide power for 25 days. For a lighter weight version a lithium coin cell can be used. After recording the SD card is removed and downloaded in a regular computer.

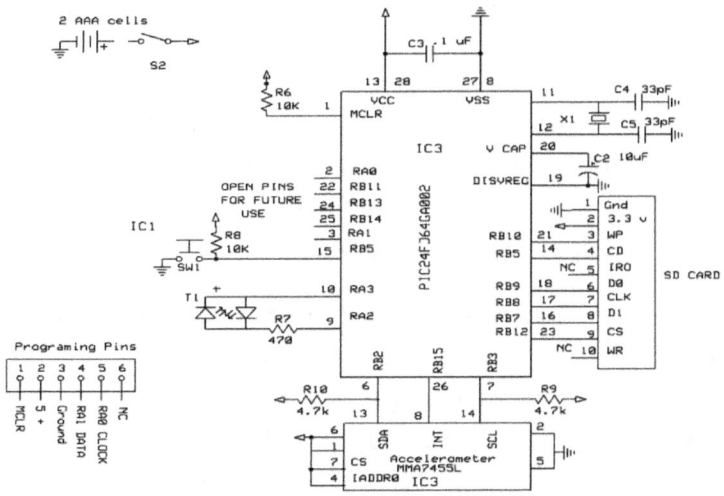

ELECTRONICS:

The microprocessor used is a Microchip 16 bit Pic24J64GA200. Programming pins are made available on the board for use of a PicKit 2 or 3 programmers. The software is free from Microchip. To change the programming you need to know how to program using C language. There is plenty of documentation on how to interface with SD cards with this chip. (See N&V October 2010 "Implementing a File I/O for the 16 bit Micro Experimenter" by Thomas Kibalo.   The power for this chip is 3 volts and not 5. The accelerometer chip has a range of 2.4-3.6 volts.

 If you want to make your own boards, the Express PCB board files are on the CD along with a parts source list and Hints and Tips. I had planned to have an extra board for the accelerometer so that the reader could mount their own accelerometer to keep the cost down. After ruining several accelerometers trying to solder by hand, I decided to design the board for using the Parallax Accelerometer. However, for those who want to cut the costs and try their hand at surface mounting, the board file for the accelerometer is available on the CD. The SD card sits above microprocessor and is easy to remove. The board was designed round to fit into a 3" PVC tube.

The microprocessor talks to the accelerometer using I^2C. The accelerometer has a built in A/D converter and puts the results out in digital form. The micro is put to sleep in the seismic and shipping modes, thus it draws very little power.

 The SD card programs the micro with date, time and the mode and the tripping points. There is an Excel program that is on the CD to program the SD card. Each time an event happens the date, time and results are recorded on the SD card. The files are named Accel1.txt, Accel.txt, etc. Each event advances the name of the file up to 256 names. The led will flash every 4 seconds indicating that an event has taken place. The card can be removed any time by turning off the power and placing it into a computer for downloading and viewing.

CONSTRUCTING THE BOARD:

There are two boards for this project. The first one is for the SD card. They are available from the CD.

Turn the small board over and solder the six 10K 805 surface mount pull up resistors. The best way of doing this is to melt a small amount of solder to top six pads. Using tweezers to hold the 805s touch the tip of the soldering iron to the pad and place the 805 on the pad allowing the solder to flow to the resistor. Do this for the rest of the resistors. Now using solder, solder the other side of the resistor to the pads. Make sure you use rosin core solder. I normally use .6 mm diameter. Turn over the board and solder the SD holder to the top side. Solder each of the pins and ears in their respective holes.

The second board will need to be trimmed to the outside circle. The easiest way is to use a sander. Make sure it will slip into the 3" PVC pipe so you don't have to sand after the components are placed on the board. All the components are placed on the top of the board.

The board will have areas that are marked for components but they will be left vacant. The board was designed for two separate projects, the" Hockey Puck Accelerometer" and "Poor Man's Seismometer"

Solder IC2 to the board. Note the square hole is pin one of the chip. Place the chip in front of you with its notch pointing left. Pin one is the lower left pin. Often there is a small dimple above it. Solder IC3 to the board. Pin one has the small voltage regulator next to it. Solder the resistors R6-7-8-9 & 10, capacitors C2, C3 and the xtal. Note the polarity of the 10uF cap. Solder the two switches. Solder the led with its long lead going to the square pad. Cut two headers, one with 10 pins and the other with five pins. Solder them to the board. There will be extra pads next to the microprocessor that can be used for other projects. Take the SD circuit board and place it over the 10-pin header. It should rest on micro. Solder the 10 headers.

Pass the two 6" wires of the battery holder through the strain relief hole and screw to the terminal noting the polarity. If you are going to use the accelerometer as a seismic detector I recommend you use D batteries so you won't have to change batteries so often. D batteries should last just less than a year.

For those who bought a pre-programmed chip, skip the next paragraph.

The chip can be programmed or re-programmed repeatedly using a Microchip PicKit II. The files are zipped and are on the CD. The programming was written in C. The square pad is Pin one for the programmer. There are only five pins needed (the sixth is redundant).

CONTAINER:

The container is made out of a 1.2" piece of 3" PVC pipe. Both the top and the bottom of the pipe should be smooth. .

Go to the CD and download the files, which go to this project. There you will find a template for the Plexiglas lens. Cut out and using a glue stick glue it to a ¼" piece of Plexiglas. Sand the Plexiglas to the edge of the circle. Using a 9/64 drill, drill three holes. Remove the template using hot water. Place the lens over the smoothed end of the PVC pipe. Mark the pipe where the holes are. Using a #36 drill, drill the three holes ½" deep. Tap each hole with a 6-32 tap. Using three 1" screws add a drop of super glue to each hole and screw each screw into the PVC pipe. Use a cut off tool and remove the heads. Glue a second piece of Plexiglas on the bottom. The top lens is secured using three knurled nuts.

Cut a piece of .06" x .125" Styrene plastic strip to 7.75 inches. Bend the strip and place 5/8" from the edge of the PVC pipe. Using super glue, glue this strip in place. This will act as a support for the board.

USING THE UNIT:

The modes and several parameters of the accelerometer are set by programming the SD card using an Excel spread sheet.

Using your computer*, format a SD card from 256 MB to up to 2.0 GB. A seismic recording only takes up 35.1 kb. That's many event recordings. Now we are going to set the time in the microprocessor's clock. **NOTE: If the power is turned off, the micro's, clock will be lost.** From the CD, download the Excel sheet titled "Accelerometer Programming". Follow the directions in the spreadsheet. The time is written to the SD card using a BCD format in hexadecimal code. Fourteen six bytes of code in one long statement are stored in time.txt. Make sure you pre-program the time ahead to allow enough time to place the card into the unit.

SEISMIC DETECTOR:

When in the Excel Spread Sheet, type in an x next to Seismograph. Load the card. Turn on the unit. The led should be red. Watch the clock and when it reaches the proper minute, press switch one and release. The led should flash green indicating that the micro's clock has been set to the time you programmed on the SD card.

Once you have located the place you are going place the seismograph, you can activate it by pushing the switch. The led will turn on steady red for one minute. This will allow you to place the lens on and secure it with the knurled nuts. When the red led goes out, it indicates that it is armed.

When the unit is jarred or an earthquake happens, the data will be written to the SD card and the led will turn a green indicating that it is making a measurement. Once the unit has made a one-minute measurement, the led will flash red about every 4 seconds indicating that an earthquake has occurred.

When you are ready to read the data, remove the lens (this will activate the unit and record the jarring of the removal also but will end up as an invalid txt file). Turn off the power and remove the SD chip. The "Seismograph" has fixed parameters and cannot be changed using the Excel spread sheet.

CONTINUOUS MONITORING:

When in the Excel Spread Sheet, type in an x next to Continuous Monitor. Load the card. Turn on the unit. The led should be red. Watch the clock and when it reaches the proper minute, press switch one and release. The led should flash green indicating that the micro's clock has been set to the time you programmed on the SD card. Press the switch and the led should be a steady green indicating that it is collecting data. To stop recording, press the switch again. The led should not be lit. Turn off the power switch and remove the SD card. The g ranges can be changed to 2g, 4g, or 8g, using the Excel spread sheet.

SHIPPING MONITORING:

When in the Excel Spread Sheet, type in an x next to Shipping Monitor. You will need to type in the trip point you want the unit to start recording. This can be 0 .1-8 g

Load the card. Turn on the unit. The led should be red. Watch the clock and when it reaches the proper minute, press switch one and release. The led should flash green indicating that the micro's clock has been set to the time you programmed on the SD card.

Once you have located the place you are going place the shipping monitor, you can activate it by pushing the switch. The led will turn on steady red for one minute. This will allow you to place the lens on and secure it with the knurled nuts. When the red led goes out, it indicates that it is armed.

When the unit is dropped, jarred or thrown, and exceeds its trip point, it will record for 5 seconds (this can be changed) when an event happens, the data will be written to the SD. Once the unit has made a measurement, the led will flash red about every 4 seconds indicating that the trip point has been exceeded.

There are also several parameters that can be changed. Motion detection, Free Fall detection, Integer trip value, absolute mode, recording time and g range.

NOTE: Only 1 mode can be used at a time.

VIEWING THE DATA:
To view the data, go to www.dataq.com and download their free software under downloads titled "WinDaq Waveform Browser ". For those that are interested in great data loggers and data acquisition, I can't say enough good things about "WinDaq". The browser is also located on the CD.

Once the "Waveform Browser" is downloaded, there should be an Icon on your computer. Click this icon. A window should open up. Locate the SD card and make sure you have "Files of type" on "(*.txt)", look for "Accel1.txt" file or other numbered seismic files. Click on this Icon. Perform a save Accel1.wdc (default). Use #2 "Spreadsheet print file (ASCII)". The convert screen should pop up and default on "volts". Hit the return key. (Subject to change depending what version, you have with the Browser.)

The Waveform Browser should now be visible. If you go to "options" and click on "Cursor time", the date and time will show at the bottom of the time the earthquake happened. Moving the cursor below the red line, will show the time. By pressing F4 to place a time marker you can measure from the time marker the seconds from an event.

There is a plethora of commands you can use in the "WinDaq" browser, and I would defer you to their help menu, as it would take a small book to explain them all. Once you have down loaded your data, delete all the Accel.txt files on the SD card to re-record.

CALIBRATION:

The unit must be calibrated as each sensor is different. Calibration method is located in the Excel spreadsheet.

Happy egg dropping!

HOCKEY PUCK ACCELEROMETER

CHAPTER TEN

LVDT
LINEAR VARIABLE DISPLACEMENT TRANSDUCER & HOW TO MEASURE THE EARTH TIDES

August 30, 2012

Ron Newton

This project is three projects in one. It is a device that will measure the Earth Tides, not the ocean but the tides of the semi-solid structure of the earth. It is also very sensitive Seismograph and Horizontal Gravimeter. It can be built for less than $30.00 and does not use a microprocessor. You will need a chart recorder or I recommend using an old computer and a Dataq DI-145 Data Acquisition module which is a must for any experimenters tool box. It retails for $29.00.

EARTH TIDES:

Carson City, Nevada sits on 25 earthquake faults and I'm directly over one. The Sierra is still growing so we constantly get small shakers. In 2003 I set out to find if there was any relationship between the earth tide and earthquakes. I designed a tiltometer using mercury pools. It was extreamly sensitive and could measure a tilt factor of a dime placed 100 miles away. Unfortunately mercury is hard to come by and has received a bad name. One of local schools was shut down for a week as a student had dropped a medical thermometer and the local Hazmat came in with respirators and white suits to clean it up. In my younger days, we used pints of mercury in the physics and medical labs and it was always all over of the floor. Most of us in my age group can remember rubbing pennies with mercury to make them shiny.

The earth is a big grapefruit being squeezed by the moon. Most people understand that the ocean has tides but don't realize that the earth changes diameter as its core is liquid (see Measuring the Earth Tides included on the Nuts and Volts website). After 8 years of continuous recording, I determined there was no relationship.

This device with a two foot tube should be equivalent to the mecurical one.

SEISMOMETER:

This instrument with a two foot tube should be able to measure an earthquake anywhere in the world with an earthquake of 6 or more on the Richter scale. For the above experiment I had made a seismograph that had appeared in Scientific American which had only one dimension. This one is two dimensional.

HORIZONTAL GRAVIMETER:

Actually this project started out as a Horizontal Gravimeter but for the reader it is far more popular as a Tiltometer or sensitive Seismograph.

In preparation for the solar eclipse coming in November 2013, there are some very interesting questions of what happens to gravity when the moon, sun and earth align. Abnormal readings have been made using vertical gravimeter. Research was made by the students from Norway on the horizontal components of gravity with no conclusions. One of the difficulties with measuring the X-Y coordinates is the tilt factor. The vertical G meter was made using conductive foam which worked well, however, I needed to a more sensitive instrument. The second dimension and the foam did not lend itself for this project. Using a heavy pendulum and measuring the change of capacitance seemed the way to go.

THE UNIT:

The capacitor is made of the pendulum and four surrounding plates facing north-south, east-west.

It was determined that the theoretical capacitance would be in the area of 16 pF. Measuring capacitance in the pF mode is challenging as using a capacitance meter as its lead capacitance becomes problem.

To the rescue is the LVDT using a Wheatstone bridge. The Wheatstone bridge is configured so that when the pendulum leans to the North the capacitance increases and the South capacitance decreases. When the difference between them is taken, it helps cancel out noise and give a much more practical voltage then just trying to measure one capacitor. The Wheatstone bridge when using only resistors uses a dc current and is easy to measure with a dc micro-voltmeter. When all 4 resistors are equal there is no voltage. You can also use an ac current for a signal and I can remember in the 1940s using an ac signal in the audio range and headphones listening for the tone to disappear and determining the null point.

ELECTRONICS:

The LVDT is a great circuit for number of projects. Even if you don't want to build this project in Toto, the board has many applications of measurement of other project using resistors, capacitors, and inductors. The circuit uses a Wheatstone bridge which can be extremely accurate. It was invented in 1842 by Sir Charles Wheatstone. Several modifications have been made throughout the years with the application of ac currents. Here is the dc version.

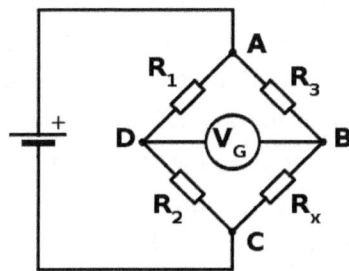

Without going through a lot of math if R1, R2, R3 and Rx are all the same no voltage will be available across D and B. If however R1, R2, R3 are all the same and Rx is lower than the others a voltage will occur across B and D with B being negative and D being positive. If Rx is higher than the others D will becomes negative and B will be positive. Now comes the magic, If R2 and Rx become a potentiometer with a resistance of 200,000 ohms, and R2 was at the higher resistance than Rx, e.g., 101,000 ohms, Rx must be at a lower resistance of 99,000 ohms (200,000-101,000 = 99,000), now even a higher voltage occurs between D and B. D being positive.

 Now the problem becomes we are using a capacitor and not a resistor. How can we turn a capacitor into a resistor? If ac is placed across a capacitor, it will act as a resistor but we change the terms from resistance to reactance. Both are in ohms. The reactance in ohms is directly related to the frequency. It was determined by both imperial calculation and actual measurement that when the pendulum hangs in the center of the copper plates the capacitance is 16 pF. Using the formula for capacitive reactance vs. frequency,

$$X_C = \frac{1}{\omega C} = \frac{1}{2\pi f C}$$

Therefore at 100,000 Hz the reactance in ohms is about 100,000 ohms. If R1, R2, R2 & Rx = 100,000 ohms, D and B are both = at 2.5 volts with no difference between them. However, any change of the pendulum from seismic, horizontal G forces or tilting will unbalance the capacitance and an ac voltage will change across D and B. A differential instrumentation operational amplifier was used as it eliminates a lot of noise due to common mode rejection ratio (CMRR). It amplifies the signal by 10 which was determined by experimentation. This signal ac is converted by half wave a precision rectifier using a LM741. The output is filtered by C1 and C2 now can be charted by a chart recorder or a Dataq accusation module.

The only stumbling block with this project is that you need a lathe or know someone who has a lathe. The sensor fits into a 1.5" PVC coupling. The coupling needs to be turned so that the 1.25" copper coupling fit into it. Tolerance is not an issue as long as the copper coupling slides in. The last 1/4" is left unturned so that it provides a lip to hold the copper capacitor plates. Sand the outside of the coupling first to make sure to get the wrench knobs off so that it will slide into the 2" PVC tube. Once this is done the 1.25" copper is split into 4 parts using a hack saw. There is a template on the CD for this. Use glue stick or rubber cement to hold the template on the outside of the coupling. Using a sharpie, mark the inside of the coupling 1-2-3-4 in the different sections. This will insure that if the hacksaw cuts are not parallel that the pieces will still fit together. After cutting, smooth the edges of the copper. Remove the template. Solder four 6" wire wrap wire to the top outside of each of the pieces with the end of the wires going downwards. Two blue wires on pieces on 1 & 3 and two white wires on 2 & 4. Take piece 1 and using super glue, glue it into the PVC collar. Use the back of a hacksaw blade or feeler gauge for a spacer and glue #2 next to it. Do the same for 2 & 4. Using an ohmmeter, check to make sure there is not a short between any of the parts due to filings. The final project should resemble a commutator. Mark the end of W & S wires with black. Now wrap the wires from N, E, and W around the copper plates until they meet S and twist together. Tack them to the copper using super glue.

BASE PLATE AND TUBE:

The base plate was made out of a plastic kitchen cutting board purchased for less than $6. Use the template for cutting and drilling. It can be cut either in a circle or a triangle. Tap the leveling holes with a 6-32 tap. Tap the 3 holes for the level with a 4-40 tap. Deburr all the holes top and bottom.

If using it for a "Tiltometer or a "Seismograph use a 24" x 2" PVC pipe. If used for a "Horizontal Gravimeter" uses a 12" piece. The PVC pipe needs to be cut square. Using a #43 drill, drill a 1/2" in depth on the end of the tube and tap using a 4-40 tap. Mark this hole and the hole on the bottom of the base plate with a black marker. Secure the tube to the base plate using a 4-40 screw. Make sure the tube is centered. Now turn the unit upside down and drill the other two holes into the tube. Remove from the base plate and tap both holes. Drill a 3/16" hole 2.3" from the bottom of the tube in line with the hole with the black mark. Thread a red wire wrap wire into the hole from the outside and up the inside of the tube leaving about 3" of wire coming out of the end of the tube. Tack the wire to the inside of the tube using super glue.

Screw on the "Bull's eye surface level".

Using the copper assembly feed the wires into the hole and slide the assembly into the bottom of the 12" pipe so that the assembly is even with the bottom. Make sure the wires are not on the inside of the assembly. Screw three 1" 6-32" leveling screws from the top of the base plate so they extend 1/2" from the plate. Add three acorn nuts to the screws and put a drop of super glue on each.

Take the 1.5" end cap and drill 7/64" hole in the center.

PENDULUM:

The pendulum is made out of copper. Cut a 3" piece of 1" copper tubing and deburr. Make sure that 1" sleeve sides on to the tube. Couplings have stops in the middle; make sure you pick up a sleeve. Clean all the copper parts using a fine sand paper to remove any oxidation.   Slide the 1" coupling on the tube and center it. Using a torch, tack the coupling to the tube using flux and plumbing solder. Take an end cap and solder it to the pipe flush to the coupling. You don't have to solder all around only enough to hold it. Take the other end cap and drill in the center a 7/64" hole. This must be in the center. Using .5" 7/64" aluminum ball chain, add a clip to the end ball. Feed the chain from the inside of the cap and add a second clip. Fill the tube with lead shot or sand. **DO NOT USE STEEL SHOT AS IT IS MAGNETIC!** Place the end cap and solder. Turn the pendulum on a lathe so its diameter is 1.2". This will give plenty of distance for centering the pendulum.   Add the rest of the chain and feed through the 2" PVC cap.   Using a 1/2" spacer of some type, place in the bottom of the tube on the table.   Drop the pendulum down the 12" tube allowing it to sit on the spacer. Put the end cap onto the pipe. Lift up the chain and mark the last ball that is even with the end cap. Turn the tube on its side and pull out about 1" of chain. Cut the ball chain at the mark and add a clip. Now strip about 1" of the red wire and wrap it around the aluminum chain close to the end cap. Put on the end cap. Using three 4-40 screws align the black marked holes together and secure the tube to the base plate. Mark the leveling screw opposite the wire hole as North or leveling screw one.

BUILDING THE BOARD:

The Board files from Express PCB along with "Hints and Tips" are on the CD.

Solder in the five IC chips. Pin one goes to the square pads. Solder in the four diodes and the 1 uF capacitors. Note their polarity. Solder in the resistors and the switch. Cut two pairs, a set of three and one pin from the header strip. Solder them into the large pads, single to pendulum the pairs to N-S and E-W and triplet to the three pads on the lower right.

Thread the 9 volt battery snap wire leads through the strain hole. From one snap solder the red wire to + 9V and black to the GND pad. Solder the other snap black lead to the -9V and the red to the GND pad. Thread the 6 volt AAA battery pack into the strain hole and solder the red to the 6+ and black to the Gnd by pin 8 of the PIC12F508. **Note:** You can use up to ± 18 volt power supply in place of 9 volt batteries.

Using the 1/4" spacers and 4 1/2" 4-40 screws, screw the board to the base plate.

Using a wire wrap tool, wrap the red wire to the pendulum pad and the four remaining wires to their respective posts.

SETTING UP:

Place the unit on a level table. Using the "Bull's eye surface level", level the base plate. Note the north screw as one and going clockwise the next screw as 2 and the third as three. Now take an ohm meter and clip it to the clip on top of the end cap. Place the other lead on the east wire coming from the copper plates. The resistance should be open. Do the same for the west plate. If not, adjust leveling screw two until both readings are open. Do the same for the north and south plate. Use screw one leveling screw for this adjustment. If turned, make sure to recheck the east and west.

Turn on the 6 volt supply using the switch on the box. Turn on the dual voltage supply using the switch on the board.

Connect the board to the Dataq accusation module or a chart recorder. North should be on channel one. Make sure your computer has the right data and time. Open the Dataq program open "Edit" and click "Channel" and make sure you are reading two channels. Go to" view" and use the one page both channels. Press F3 and set for 100 readings per second. The top chart will be north and south, north being the higher voltage and the lower chart will be east and west with east being the higher voltage. When the unit settles down, adjust the voltages to about 1 volt using the leveling screws. Keep in mind that we are using a differential amplifier and the ac voltage is rectified. If the pendulum is in the center the voltage output will be zero. A pendulum displacement on either side of center will produce a positive voltage. Therefore you must provide an offset by tilting the unit a small amount. Start with east and west. Turning screw two counter- clockwise will place the pendulum nearer the east pole. If you get a pattern oscillation of 1/2 a large sine wave follow by a smaller one you are near the center point. Adjust for about 1 volt with a full sine wave. Once you get the east west set, adjust screw one for the north and south. Turning screw one counter- clockwise will increase its voltage. Adjust for about one volt.

Any vibration will cause the unit to change with the pendulum wanting to stay in it location due to its mass. Any tilting will do the same thing. Horizontal G forces will move the pendulum, while the base is fixed to the earth.

EARTH TIDES:

Once everything has quieted down, Press "F3" and set the recording to 1 per second. Click "File" and then "Record". Name the file and make sure you have enough memory for 24 hours. This can be checked by putting in a number, e.g., 250 and clicking "OK" and holding it to read the number of hours to record. It is not enough

94

release "OK", if not slide the arrow off "OK" and try again. When you have recorded all the data you want, go back to "File" and click "Close".

SEISMOGRAPH & HORIZONTAL GRAVIMETER:
Press "F3" and set the recording to 100 samples per second. You will have to increase the file size as you are recording a lot more data.

Once a file has been closed, the files can be reviewed using the Windaq Waveform Browser by clicking on the file. Using "F7" and clicking Maximum will display all the data on one page. Once in the browser you can amplify or look at smaller sections. To see what time a seismic event took place, Click on "Options" and then on "Cursor Time".

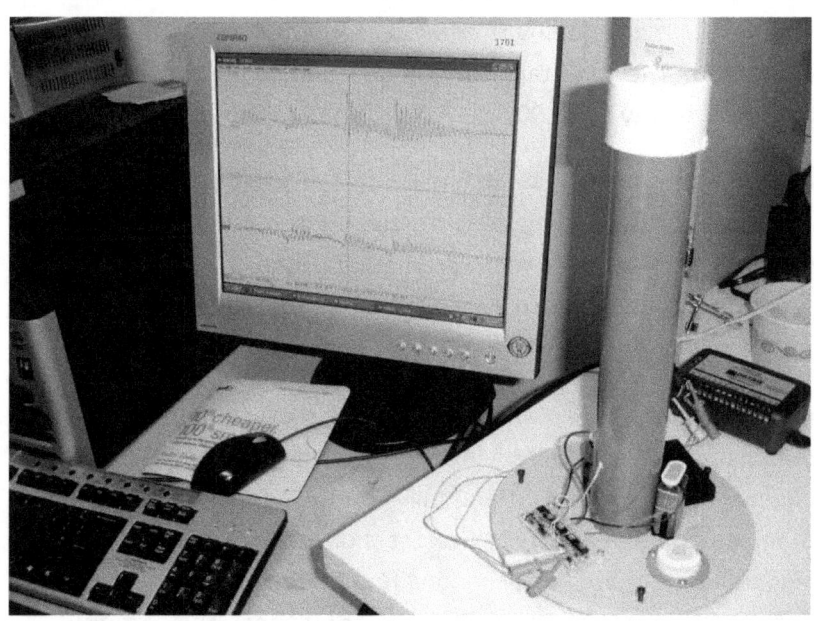

HORIZONTAL GRAVIMETER, SEISMOMETER & TILTOMETER

✱✱✱✱✱✱✱✱✱✱✱✱✱✱✱✱✱✱✱✱✱✱✱

CHAPTER ELEVEN

SORTING COUNTER

Published July 2011

If you visited a medical laboratory 25 years ago, you would have probably found a Med Tech looking though a microscope using a hand tally counter counting white cells, red cells or platelets. Or they might be using a sorting counter for differentiating white cells. Now days in the USA, most of this work is done using automated equipment. In developing third world countries, they still use the old method. The hand tally counter cost $40.00 and the sorting counters are now $600.00.

I combined both devices with a parts cost of $15.00. + Board. This is a great project for the beginner or student as it will get you started soldering and interested in programming. There is one surface mount (battery holder) and everything else is through hole. The nice thing is that it will fit in your shirt pocket.

Sorting counters have many uses, keeping score, counting parts; counting people e.g., classifying their sex, age etc, it is just a handy gadget to have on hand. Hand tally counters are often used for counting people at events or even cars on the road. Instead of using a sheet of paper and marking four lines and a slash showing a count of five you just push the button or buttons. The sorting counter allows you to keep a total plus keep track of specifics. E.g., if you are counting VWs on the road and want to know the percentage of red, blue, yellow, green and black you would use the differential counter.

This is a very simple project for those who want to learn to solder or are interested in using microprocessors and how they function. No special tools are needed beside a small tip soldering iron. It has no box as it stands alone, therefore there is no drilling. If nothing else build one and send it overseas to a third world hospital. If you want to change programming, you will need a PIC II Microchip programmer. ($34.95 www.microchip.com).

The main thrust of this article is to teach the novice about displays and multiplexing. Check out the Nuts & Volts website store. They have many great programming books for the PIC line.

WHAT IT DOES:

This project has 5 counting buttons for sorting and tallying. One mode button (blue) for changing from differential to hand tally's and one button (red) for clearing results.

When using the sorting function, it will count 100 items displaying the count, sound an alarm at 100, and then when you push the sorting buttons, it will display the percentage of each button.

When using the hand tally, it has three buttons, which will tally the counts. The first button advances the count by one. The second button will count in batches of three. The third button will count continuously with a pulse rate of 5 counts per second as long as the button is held down.

Pushing the reset button resets the counts to zero. The leds will display up to 9,999 counts.

If no input is performed for three minutes, the leds shutdown and the micro will go to sleep. Pushing the reset button (red) will wake it up. No data is lost. It runs on two 3-volt CR2035 batteries.

CONSTRUCTION:

Place 4 rubber feet on each corner of the bottom side to prevent scratching the traces on the bottom of the board. Mount the Pic16F726 with its pin one to the square pad labeled 1. (The micro is mounted upside down with pin one to the right.) Solder all the pins of the micro. Next, solder the 4 PNP transistors their flats pointing toward the top of the board. Solder the 470 ohm 8 inline resistors above the micro. Bend the leads of the four 10 K resistors at right angles and place them in their respective holes and solder. Solder the transducer and C1 noting the + going to the square pads. Solder the seven switches in their holes. The five brown switches are inline on the bottom. The red switch goes on the left and the blue switch goes on the right above the brown switches. Flood solder on round circle on the battery holder for making a contact for the negative terminal of the battery. Solder the battery holder with its opening toward the edge of the board. Solder the 4 digit led. The 6 pads will remain empty, as these are the programming pads. Clean any remaining flux and solder with alcohol.

USING THE UNIT:

Slide two CR2032 Lithium 3 volt batteries into the holder. The positive should be up. Press the mode button, the decimal of the one digit will turn on and off indicating if the unit is in the tally mode or differential mode. With the differential mode, the decimal is off. A decimal indicates the tally mode.

TALLY MODE:

Make sure the decimal is on.

1. Press key one (far left key), the counter will count once for each push.
2. Press key two, the counter will count three counts for each push
3. Press key three, the counter will continue at 5 counts per second as long as the key is pushed.

The counter can be cleared at any time by pressing the reset key.

DIFFERENTIAL MODE:

Turn off the decimal by pushing the mode switch.

1. Assign the keys to what you want to sort.
2. Press any key. The counter should increase.
3. Press any other key the counter should increase.
4. When the counter reaches 100, the alarm will sound and pressing the keys will no longer increase the count.
5. Hold down any key and it will display that key's percentage.

ELECTRONICS:

The PIC 16F726 made by Microchip has many attributes. It has a voltage range of 1.8-5.5 volts (can go to 6.5 V) When put to sleep it only draws 20 nano-amps (.00000002 amps). It does not need an external crystal and each pin can sink or source 25 milliamps. It even has 14 analog to digital converters (not used in this application). 12 interrupt capabilities, plus many more capabilities all for $2.31 ea.

The best thing I like about the F series is that they can be programmed over and over even when soldered in the circuit. I use to use programming pins but developed an adapter for the PIC 2 programmer that uses spring-loaded pins that you just touch the programming pads. (These are the pads next to the battery.)

The switches go to separate pins of the micro, which have internal pull up resistors. This eliminates having to place external pull resistors which I had do a few years back. The switches when pressed take the inputs to ground. The reset pin is placed on a special pin that is called an interrupt pin.

MULTIPLEXING:

Let's take one display at a time. Displays come as common anode and common cathode. Take a look at Figure one (see CD). With a digital display, it has 8 segments and sometimes a decimal or colon. If you are using a common anode, all the segment anodes are tied together. If you place the positive of a battery to the common terminal and touch the segment cathode to negative (via a resistor to limit the current) it will light that segment. With a common cathode all, the segment cathodes are connected together. You just reverse the polarity of the power to the chip. With the sorting unit, I chose to use a common anode. No particular reason, it was available. Connect the positive terminal to the common anode to display an "eight" ground all the segments. To display a "one", ground cathodes B & C.

Now there are four digits in the display I used. You could use a micro or several micros to provide 8 pins for each of the digits and tie all the anodes together, but this would mean you would need 64 pins for the display alone! What if we tied all the segments as, bs, cs, ds, es, fs, gs and decimals together and separate the common anodes from each digit? This would only require 8 pins for the segments, and if we connected the four anodes to pins to the micro (via a PNP transistor going to VCC), we would only need 12 pins instead of 64.

The human eye cannot detect a flicker above 50 cycles per second (50Hz). This is why the USA went to 60Hz in the 1940s from 50Hz. Europe often still uses 50Hz and you can see the flicker in fluorescent lights. (Incandescent lights don't flicker due to the filament staying hot.)

If you ground a, b, c, d, & g segments and turn on its transistor, it will display a "3". Turn off digit 1 and ground a, b & c and turn on its transistor it will display a 7, and so on. If this is done at a frequency above 50Hz, they will all appear to be on at the same time. The micro runs a 4 mega-hertz so this is not a problem.

SOFTWARE:

Download the free software MPLAB from Microchip. Go to CD and open the Sorting.ASM in MPLAB. Make sure you turn on the line numbers. You always have to keep in mind that most micros count in hexadecimal (base of 16) 0, 1, 3, 4, 5, 6, 7, 8, 9, a, b, c, d, e, f and that the displays display in decimal (base of 10) 0, 1, 2, 3, 4, 5, 6, 7, 8, 9. This means that the hexadecimal needs to be converted to decimal. E.g., F = 15. This is done by a canned program I picked up on Microchips website. Why re-invent the wheel?

Line 595 starts the Display section. All the digits are turned off.

I use a look up table for the display to display the numbers. The lines 602 to 607 insure that jump to the lookup table do not end up in oblivion. This can be a problem with lookup tables. Let's say there is a three in the ones register and we want to display it. Three is loaded into the "W" register (working register) and then a call is made to the lookup table called "PATTERN" Take a look at starting line 733. Pattern is the name of the lookup table. The ADDWF tells the micro to add the number in the W register to the Program counter. The remaining 10 lines provide the inputs and output for turning on the segments. 0 = ground and 1 = Vcc to the pins going to the leds. W was loaded with a three, 3 + the program counter = line 739. The RETLW returns the binary code (B'01001011') via the W register and places the binary code into the PORTC, which controls the leds. Digit one is then turned on. A delay is called so that it will display for a period of time. Digit one is then turned off.

The program does the same thing for the tens, hundreds, and thousands looking up each number that needs to be displayed.

SWITCH BOUNCE:

If you are not familiar with microprocessors, you probably didn't know that when most switches and relays are turned on their contacts bounce on and off several times. This is normally not a problem. However, with the micro running at 4 million times a second it will count the bounces. Fortunately, the switches in this project give very little bounce; however, even if they didn't bounce, you would have trouble releasing the switch to the off position in that amount of time and would probably get a count of several thousand before it turned off. I you take a look at the 5 lines after the label "START, you will see that the program calls for a "DEBOUNCE" which is a time delay. The program will hold until the switch is back in the off position.

CALL VERSUS GOTO:

A call is a routine that is used quite often. Instead of writing the "DEBOUNCE" (a timer delay) each time, you just call it. It does its function and returns to the next step after the call.

'GOTO" causes the program to jump to a section of the program but will not return. It is often used after a BTFSS (Bit Test File and Skip If Set) or a BTFSC (Bit Test File and Skip if Clear) to perform another function. The $-1 after a GOTO causes the program to go back one step. You can use $-5 to go back 5 steps. Be careful though, sometimes it won't jump back in the right place if you are at a page break.

For more information on labels and how the program works see "Hints and Tips" on the CD.

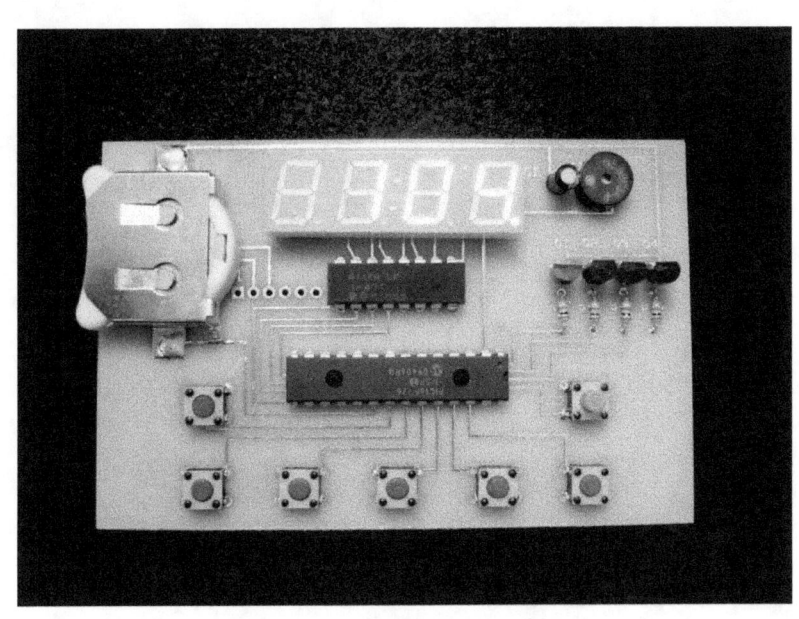

SORTING COUNTER

****

CHAPTER TWELVE

PORTABLE FIELD INCUBATOR FOR MICROBIOLOGY

By Ron Newton
Built 2004

TOOLS NEEDED
Milling Machine and lathe.

With my many travels to Developing Countries (the use of Third World is frowned upon) I'm often one of the first to arrive. My background included being a Microbiologist and my cohorts often ask me "Is the water safe to drink?" Fortunately now days there are an abundant amount of bottled water, beer or Coke. I normally drink beer as you don't know if the locals are making their own bottled water out of the tap. Beer, due to its pH, eliminates the bad bacteria and it uses yeasts for fermentation. If there is a bad bacterium in it, it will be unpalatable.

I test the tap water at different locations for coliform (the bacteria in sewage). In Mozambique, I designed a new 40,000 square foot laboratory and checked the local well. Right next to it was the local hospital with tents in the back for treating cholera patients when epidemics occur. In Africa it is not uncommon to see both men and woman squatting on the hospital property defecating as there are no public toilets. The well tested TNTC (too numerous to count) per

100mls of water, and I found Vibrio Cholerae which is the organism suspected of killing Tchaikovsky.

Millipore® makes a portable kit which uses their 47mm disposable Petri dishes. It has a syringe and flask that pulls water through a 2.2 micron sterile membrane which catches all the bacteria. This membrane is placed on a piece of thick absorbent paper. The media for growing the bacteria comes in small vials which are broken and the solution is placed on the membrane. The plates are turned upside down and placed in a 35 degree centigrade incubator. The plates are read at 24 & 48 hrs and the bacterial colonies are counted. The problem is that unless you have access to a laboratory, you need an incubator.

In the 1990s I developed a small heater for shipping platelets which have to be shipped at 20-24° C. It uses 4 C batteries. Most of us in electronics know that transistors can get very hot and need to be heat sunk. I took advantage of this by using a NPN transistor and use it as a heater by shorting its collector and emitter across the batteries. The base is driven positive by a microprocessor using pulse width modulation for temperature control. In series with the base is a current limiting resistor and led which flashing is directly proportional to the amount of heat being called for. This gives a visual indication that the unit is on.

The microprocessor uses a 10K thermistor in series with a 10k 1% resistor as a voltage divider. On this portable incubator I added a potentiometer to adjust the temperature and substituted for R1. As the temperature increases the thermistor's resistance decreases. The microprocessor measures the voltage and compares it with predetermined voltages and turns on and off the transistor. Once the temperature settles down, the micro changes the pulse width in small increments which gives very precise control.

Since the microprocessor requires 2.5 - 5 volts to function, for the incubator I used a LM7805 1.5 amp positive voltage regulator which is good up to 35 volts input. This voltage regulator also produces heat at whatever amperage the transistor is pulling to heat also appears across the regulator. E.g., if the transistor is pulling 500 milliamps at 5 volts (watts = voltage x amperage = 2.5 watts), the regulator also has to supply the 500 milliamps. If the input voltage is 12 volts, 12 - 5 = 7 volts. 7 volts @ 500 milliamps = 3.5 watts. This heat is also used to heat the incubator.

For a power supply in hotels, I use a 12 volt dc switch mode combination 110-220 volts ac converter. If I am in the field performing research on rivers and streams, I just plug the incubator into the cigarette lighter plug.

CONSTRUCTION:

The inside of the incubator is made up using two ea 2" copper sleeves used for joining 1.5" copper tubing and an end cap. Make sure you use copper sleeves and not a coupling as the couplings have a ridge inside. This assembly will house 15 Petri dishes. On the CD is a template for drilling the holes in the bottom of the cap. Cut the template and use glue stick to glue the template to the bottom of the end cap. Drill the two holes using 5/32" drill. Deburr the holes and place two 1/2" 6-32 with a neoprene O ring from the inside. Secure with two 6-32 nuts. Before soldering the tubes together, using a lathe turn the inside diameters to 2.15".

The copper sleeves are cleaned on the outside and using acid core solder and flux are soldered together using along with the end cap. I used about 12" of aluminum flashing coiled and fitted in the sleeves to center the inside. Once soldered, I turned them on a lathe to remove any excess solder on the inside. Make sure that the Petri dishes will slide down the tube.

CIRCUIT BOARD:

Place the board so that you can read the word "Incubator". Place the Pic12F675 with pin one in the square pad. You will need to program the chip or use a preprogrammed chip. (See CD) Add the two resistors and the capacitor. The long lead of the led goes to the square pad. Solder these components. Now turn the board over. We are going to mount the LM7805 and the 2SD2012 upside down. Bend the leads of the transistor at right angles opposite the flat side. Solder in place. Solder the thermistor. It has no polarity. Solder an 8" red wire to the positive pad and 8" black wire to the gnd pad.

Smear the two transistors with heat sink compound. Place a 1/4" # 6 nylon spacer under each transistor. Remove the two nuts from the bottom of the end cap and place them through the transistors and spacer secure on the other side with the two nuts. Coat the nuts and screws with finger nail polish to secure against vibration. It is important that the bottom is water tight as any leakage will damage the circuit board. I speak from experience. That is the purpose of the O rings. Wrap several turns of electrical tape around the bottom extending over the circuit board. This will prevent the insulation blocking the led in the next step. Do the same for the top of the copper but use a stiff piece of paper inside of the tube with the tape.

I use a double insulated 24 oz slurpee cup with a screw top lid I found at Wal-Mart for $3.50. It has sloping sides and measure 2.5 "at the bottom sits 7" high with a 4" lid with a hole in the top for a plastic straw. This is ideal as the incubator needs breath.

Check the height of the copper tubing and cut off any excess. Make sure to deburr.

 Drill a 7/16" hole about 1" from the top of the container. Put the jack's nut on the wires and feed the wires though hole. Solder the red wire to the center post and the black to the ground post. Tighten the nut to the container. Bend the leads so they won't touch the copper tube.

Using canned foam insulation like for caulking your house, center the copper assembly into the container and fill around it with the foam. If it goes over the top, let it dry and sand or cut it even with the top so you can put the lid on. Make sure that the copper does not sit above the lip which will block the lid.

Once potted, you can drill a 1/8" hole on the side of the cup for adjustment if necessary of R1.

To use just plug into power either the battery eliminator or a 12 volt supply. You might add a heavy marble in the bottom to help remove the Petri dishes.

PORTABLE INCUBATOR

****

CHAPTER THIRTEEN

UNIVERSAL TEMPERATURE CONTROLLER

UNIVERSAL?

110 to 220 volts; 50 Hz or 60 Hertz; heating or cooling; .1 amps to 10 amps; thermistor or thermocouple; - 40 to 140 degrees C @ ± 2%; -200 to + 1250 degrees C ± 2.5 degrees C, YES! It will do it, and it is solid state.

HISTORY:

Most of my dealings in Africa are with medical laboratories and blood banks. Blood particularly needs to be stored from 2 – 8 degrees C. Developing countries often use domestic refrigerators for storage that have mechanical thermostats which do not have tight enough control. Incubators (bacteriology), heat blocks and water baths are notorious for their temperature controlling unit's failure due to corrosion and lack of cleaning. Bio-hazard incinerators (high temperature) have temperature controllers that are also a nemesis. There is a plethora of manufacturers of equipment and inability to obtain parts. This is due to most of the equipment being antiquated. This brought me to design new temperature controllers for a wide range of equipment both heating and cooling. Five years ago I developed a temperature controller that would both heat or cool. I would pre-program a dip switch for most of the temperatures I expected to encounter in the field. Microchip came out with a PICKIT 2 which allows programming at any temperature in the field as it fits in your shirt pocket.

PROJECT:

This project opens up a whole new world to the experimenter, scientist or for the replacement of temperature controls of refrigerators, ovens, heating systems, air-conditioners, muffles, incubators, aquariums hot houses and so on. The cost including the box is about $40.00. It is a very simple project and the only part that may be difficult is drilling or punching the chassis. **You will need a PICKIT 2 to program it for different temperatures.** Please go to the CD for "Hints and Tips", programming files and other downloads. On the CD are Expresspcb board files for making the board or you can purchase single boards from the author.

THERMISTORS VERSUS THERMOCOUPLES:

There are advantages and disadvantages to both types. Thermistors are variable resistors whereas thermocouples generate a voltage.

Thermistors are non-linear and follow an exponential curve known as "Steinhart and Hart" interpolation laws.

$$R(T)\text{-}R_{ref} \; x \; e^{(A+ B/T + B/T^2 + C/T^3)}$$

$$T(R) = (A_1 + B_1 \ln\frac{R}{R_{ref}} + C_1 ln^2 \frac{R}{R_{ref}} + D_1 \, ln^3 \frac{R}{R_{ref}})^{-1}$$

The letters indicate thermistor factors from the manufacturer. Don't worry I have done all your work for you in a spread sheet.

The NTC type thermistors resistance goes down as the temperature goes up. (See curve 1.) They are small, come in many resistances, mounts, percent tolerances, and can be mounted and sealed using wire-wrap wire. This gives the advantage of being water immersible. They have a limited temperature range -40° C to 140° C.

Thermocouples are pretty much linear except at extremes and have an accuracy of ± 2.2°C. At zero degrees centigrade, all thermocouples generate zero volts. This is because they are standardized against the same type of thermocouple at zero degrees (ice point). This correction is made using an Analog Device AD597. Below zero degrees centigrade, thermocouples generate a negative voltage. They are great for very high temperatures but cannot be immersed directly in water and are stiff to work with. They generate their EMF using two dissimilar metals. The K type is made of nickel-chromium vs. nickel-aluminum and generates approximately .042 mv/degree.   There are many types of thermocouples, K being the most popular. They require special connectors to prevent the generation of voltages due to another dissimilar metal. They are more expensive than thermistors. The calculations for the use of thermocouples are also on a spread sheet.

PROJECT:

(See Schematic on CD.) The first half of the circuit is very straight forward. Only the transformer is fused and not the triac. The transformer is both 110 and 220 volts and jumpers are used to set the input voltage. Its secondary output is 6.3 volts. The fuse is placed before the transformer to protect it just in case the 240 volts is placed across the terminals and it is jumped for 110. Any short on the secondary side will cause the fuse to blow. The ac voltage is rectified by using a full bridge rectifier and filtered using a 470 uF capacitor. A 100 ma 7805L voltage regulator is used to provide a 5 volts controlled voltage to the components. A .1uF cap smoothes the 5 volts.

A Microchip PIC16F675 is a 10 bit A/D converter and takes the analog input signal and converts it to digital. 10 bits will provide .0049 volts per bit. 2^{10} = 1,024 bits.

$$ {5\ volts}/{1024\ bits} = .004883\ volts/bit $$

The output from the microcontroller goes to a light activated triac which controls 110 or 220 volts for turning on or off equipment. The triac specified is zero crossing which helps prevents generating radio frequencies. The project is housed in a 4" x 2.2" x 1.6" aluminum box which is used as a heat sink for the triac. There is a bi-colored led that indicates if the triac is on or off. (Red, triac is providing power, green, triac is off.)

The unit was designed for field programming of temperatures for heating or cooling (e.g., driving a compressor).

If a thermistor is going to be used, R1 is used as a voltage divider. The thermistor is used in series with the 10K resistor. The thermistor is connected to ground and the 10k resistor that goes to V+. The microcontroller compares the voltage between the two and turns on or off the triac. The amplifier is not used.

If a thermocouple is used, R1 is bypassed. The thermocouple requires a special connector for its input. An Analog Device AD 597 is single supply operational amplifier used for the amplification of the thermocouple input. It has ice point compensation built in and provides internal amplification of 245.5 V/V. This equates to 10 mv per degree C. R2 and R3 provide a divide by two voltage divider to increase the range to 1200 degrees C.

With this amplification it will give you a sensitivity of about 1 degree C. ± 2.2 degrees. For colder temperatures you will need to reverse the leads from the special connector to the board. The connector is used for both the input of a thermocouple or a thermistor for convenience. There are two jumpers inside that need to be programmed depending on which sensor is going to be used.

CONSTRUCTION OF THE BOARD:

Determine what voltage you are going to be using, and using a # 22 solid wire, short the pads labeled 220 or 110. Use one wire for 220 and 2 wires for 110. Place these jumpers on the bottom of the board for easy removal in case you want to change voltages. This way you won't have to remove the transformer.

1. Solder IC 1 and IC2 noting pins one.
2. Solder the bridge diode, 470 uF capacitor (note the +), the 7805L with its flat pointing towards the edge of the board and the .1 uF cap next to the voltage regulator.
3. Solder R1- R5 to the board. R4 & R5 limit the current to the leds. R2-R3 are the voltage dividing resistors.
4. Slide the two heavy connectors together and solder them to the board area labeled "T1 T2 T3 T4". The wire port should point toward the edge of the board. Cut two, two inch #24 wires and solder them to the two pads labeled input.
5. The headers come in strips of 36. Cut off 6 headers of the right angle and solder to the programming headers pads on the board.
6. Cut two 3 piece straight headers. Solder one to J1-J2 pads and the other to the J3-J4 pads.

See Figure 1 for board.

CHASSIS:
1. Go to the CD and download "Templates".
2. Cut out the templates and drill the holes indicated.
3. Cut the hole for the isothermal jack. This requires a square hole. (See Hints & Tips on CD.)
4. Mount the jack using the clip mounting the terminal toward the opening.
5. Place 4 ¼" grommets in the drilled area. (You will need to trim with a razor blade.)

Note: You can't use a plastic box as the triac must be heat sunk.

FINAL ASSEMBLY:

1. To mount the triac, bend the leads of the triac at right angles upward where they thicken. Push the leads from the bottom of the board. Pass the two ½" 6-32 mounting screws through the chassis and add the standoffs. Using a 4-40 screw, push it through the chassis and through the hole of the triac. You may have to adjust the leads.
2. Solder the four triac leads. The triac will provide support for its end of the board and the board at the triac end should be the same height as the standoff height.
3. Remove the board from the chassis.
4. Solder the transformer to the top of the board. Clip off any leads that extend from the bottom of the board including the terminal pins. (You may want to use fish paper under the high voltage pins glued to the chassis.)
5. Solder the bi-colored led to the back of the board. The long leg should go to the square pad.
6. Smear the bottom of the triac with heat sink compound. Place two ½" 6-32 screws through the holes in the chassis. Add the two standoffs, board and secure with two nuts. Use a 4-40 screw and secure the triac with a nut.
7. Screw the two wires coming from the isothermal terminals from the input pads. 1 goes to negative, 2 to positive. These are the two wires you would have to be reversed if using the thermocouple below zero degrees.

PROGRAMMING:

You will need to know something about Microchip programming and will need a PICKIT 2 programmer. The assembly file is located on the CD. Open it with MPLAB. MPLAB can be downloaded free from www.microchip.com.

The software that runs the controller is very simple. It just measures the voltage and compares the A/D register in hexadecimal to that of the hexadecimal preset values that you put in. This is done by subtracting one register from the other and checking the status register to see if a borrow was needed.

```
Movf   HIGH_TEMP_UPPER, 0  ; Moving user program register into
W register
Subwf  ADRESH,0             ; Subtract ADRESH from the W
register (from A/D converter)
Btfss  3,0                  ;Check status to see if subtraction
resulted in a borrow
Goto   Turnoff              ;If borrow go to Turnoff otherwise
skip next command
```

This is done for the upper byte and again for the lower byte. For checking for below a certain temperature the registers are reversed.

With a heater if the sensor voltage is above the preset it turns off, if below it turns on. Refrigerators do the opposite as their compressors need to be turned on if too warm and off if too cold. In the software you either put a 1 for heating or a zero for cooling for the mode. If you are **not** running a motor or fan, you can set the turn on and the turn off at the same temperature and not worry about chatter. The high and low presets are identical. However, if you are running a motor or compressor you need to add a differential. This is done simply by setting the turn on and turning off at temperature at different levels.

To set a differential the program uses a flag which is set or cleared. Assume the turn on temperature was set for 3 degrees and the turn off is set for 8 degrees, when turned on it stays on until it reaches 8 degrees. When turned off it stays off until it reaches 3 degrees.

This will prevent the compressor from cycling on and off at short intervals.

The bi-colored led changes its color by reversing the polarity of the pins to which it is connected.

Determine if you are going to use a thermistor or a thermocouple. Open the spread sheet for the appropriate one and put in the temperature for the thermistor and use the K look up table for the thermocouple and put in the ma generated. The spreadsheet will calculate the hexadecimal numbers you will need for the software program. The PIC F675 is a 10 bit A/D and uses two registers, one for the high byte (ADRESH) and another for the low byte (ADRESL). You will have to input both of these numbers for both the high temperature, low temperature + two modes (6 numbers). Once you have input these numbers, "Quick Build" to get the proper hex file. You can drive the PICKIT 2 programmer directly from MPLAB for this device. **Remove the jumpers before programming.** Place the programmer on the 6 header pins and program. The PICF675 can be programmed over and over.

USING A THERMISTOR:

To build a thermistor sensor, I cut wire wrap to the length I need. Wire wrap the thermistor next to its epoxy body, solder the leads and cut them short. I use a 3/8" # 6 nylon spacer and place the epoxy body in the center and fill with hot glue. Connect the wire to a thermocouple plug. A thermistor has no polarity. See Picture 2 showing several thermistors and mounting. Place a shorting bar on J1 and J3

USING A THERMOCOUPLE:

There is an alphabet soup of thermocouples. K is the most popular. This project is made for K type thermocouples. You can find bargains on E-Bay. Accessories can be found on www.omega.com . The wire is sold by the spool and comes in many coverings Place a shorting bar on J2 and J4. See Picture 3 for thermocouples.

USING THE TEMPERATURE CONTROLLER:

Power for the unit is applied to the terminal labeled "T4". The switched power is obtained from the terminal labeled "T1". Terminals 2 & 3 are the neutrals. To ground the box, use one of the two 6-32 chassis screws. I would label the grommets as S N N P for Switched, Neutral, Neutral and Power.

OTHER APPLICATIONS:

You not only can use it as a temperature controller but it also can be used as an alarm by connecting its output to a buzzer, light, horn, etc. Need a programmable temperature controller for ramping up or ramping down temperature of ovens? How about a temperature controller for a reflow oven for surface mounting? Just put a three prong plug and receptacle and plug in your toaster. The programs for both lead solder and lead free solder are also available. See Hints and Tips.

Need to lose a few calories? Just add ice.

UNIVERSAL TEMPERATURE CONTROLLER
THERMISTOR & THERMOCOUPLE

CHAPTER FOURTEEN

WATER DETECTOR ALARM AND RELAY

By Ron Newton

February 28, 2009

PROBLEM:

Plate washers for the determination of viral markers have a waste bottle for used wash fluids. The bottle should be emptied at the end of day but often it is not. The waste bottles are connected to a vacuum pump to provide suction. If the fluids are not dumped, they will find their way into the vacuum pump and corrode the valves or pump mechanisms.

Second problem: In the surgery suite, a suction bottle is used to suction out body fluids. This is usually a large bottle that goes to a suction pump. Quite often this bottle and pump are located beneath the surgery table and one cannot observe an over flow. Once the fluids over flow they often ruin the pump and it has to be discarded. A very expensive oversight.

SOLUTION:

 The February 2009 class in Kenya presented this problem. At first I presented a microprocessor solution to the problem. The second solution was to use a transistor.

Due to unavailability of microprocessors and programmers in Africa I have decided to go with the transistor model.

A NPN transistor is used with its emitter tied to ground. A nine volt battery is used for power. Two 10K resistors bias the transistor and are connected to two stainless steel wires which when contact water cause the transistor to conduct. The two stainless steel wires are pushed through the bottle stoppers and **should be below the suction outlet**.

One board was designed for both applications.

CONSTRUCTION:

All letters are on the top side of the board. The holes are plated through; therefore all components can be soldered from the bottom side.

1. If using the board as an alarm, note the polarity of the transducer and solder the transducer in place of the relay. Ignore the two pads on the top of the board. Solder in the NPN transistor pin one being the left pin with the flat facing you.
2. If using the board as a relay, solder the relay to the board. The two pads on the top go to the pump. This is one amp relay. If your motor pulls more than one amp you will need to add another heavy duty relay. Solder in the NPN transistor pin one being the left pin with the flat facing you.
3. Solder in the two 10K resistors.
4. Solder the red wire from the 9 volt battery snap to the + hole.
5. Solder the black wire from the 9 volt battery snap to the − hole.
6. Two stainless pins are used for the liquid sensor. You cannot solder Stainless steel easily, so use crimps to attach about 10 cms of flexible wire.
7. Solder the wire coming from the stainless steel wires to the holes marked "P".
8. The stainless steel wires go through the stoppers or lids of the suction bottles. Seal the wires with

Silicon glue. Make sure they are below the suction outlet.

Plug in the battery and you are ready to go.

SURGERY SUCTION:

A relay is tied between V+ and the collector of the transistor. This is a normally closed relay that is placed in series with the vacuum pump power supply. When a liquid is detected the relay activates and opens the circuit thus turning off the vacuum pump.

WASHER BOTTLE ALARM:

An alarm is tied between the V+ and the collector of the transistor. When water is detected the transistor conducts and the alarm is sounded. A Radio Shack self contained alarm was used.

The unit draws no power until it conducts and therefore has no switch.

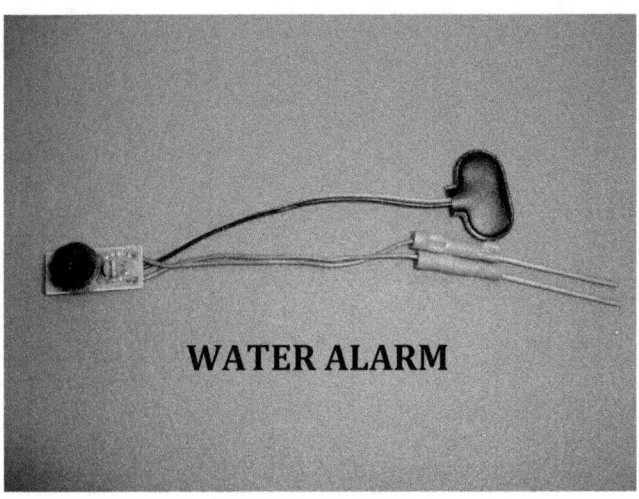

WATER ALARM

CHAPTER FIFTEEN

PRECISION TEMPERATURE CONTROLLER

By Ron Newton
03/2/2006

PROBLEM:

Many of the developing countries use domestic refrigerators or freezers. Often water baths do not have adequate temperature controls. This board was designed to bridge this gap and give precision control of temperatures. The existing controls of the units must be disconnected and the board placed between the power supply and the compressor or heater of the units.

The Temperature Controller was preprogrammed for controlling the following temperatures:

-12 to -16 C Freezers

3-5 C   Blood Bank Refrigerators

35 C ± .1 C Water Bath or air bath

37 C ± .1 C Water Bath or air bath

The unit uses a zero crossing triac capable of sourcing 8 amps @ 220 volts if heat sunk. A four dip switch is used for programming the microprocessor. The following are the programming codes:

Freezers   1 on 2-3-4 off
Blood Bank   2 on 1-3-4 off
Water Bath 35 3 on 1-2-4 off
Water Bath 37 4 on 1-2-3 off

1The temperature is measured using a thermistor and a precision 10K ohm resistor. The thermistor is connected to the board using the blue terminal. There is no wire polarity. For refrigerators and freezers, I recommend that the thermistor be placed in a solution of about 250 mls of glycerin or ethylene glycol to control hysteresis.

The power is connected to the first two right terminals (220 volts). The output is connected to the two left terminals. The board is capable of sourcing 8 amps (440 watts or volt-amps). If above this, use a heavy relay.

he board is protected via a fuse and a Varistor. If the voltage exceeds a 300 volt spike, it will blow the .5 amp fuse. A LED indicates when the unit is on. Green indicates that there is power to the board but no power to the triac. Red indicates that there is power to the board and the power to the triac.

The microprocessor is of the F series and can be reprogrammed for any temperature. Please contact me for reprogramming or the source code. With the 4 dip switch, 16 different temperatures can be programmed.

CIRCUIT:

The power can be 110 or 220 volts ac 50-60 Hz. It is protected using a .5 amp fuse in series with the power and a Transient Surge Protector (ZNR) in parallel with the power. This device has high impedance at normal voltages. When the voltage spikes its impedance drops less than 100 ohms and causes the fuse to blow. This device is often taken for a capacitor. The transformer is a split power supply and the 110 – 220 is determined wiring the jumpers on the board. The 6.3 volt output of this transformer is converted to dc using a bridge diode. This voltage is filtered using a 1,000 uF capacitor. This output is regulated using an lm7805 voltage regulator and filtered using a .01uF capacitor.

The temperature is measured using a 10K ohm (25°C) thermistor in series with a 10K 1% resistor. This acts as a voltage divider. Thermistors are not linear therefore conversions of the voltage produced must be calculated using a logarithmic formula or linear regression at 5 degree increments. The output of this voltage divider is converted in the microprocessor using an A/D converter. This particular microprocessor provides 10 bit conversions or 1024 bits. 5 volts / 1024 = .0048 volts per bit equivalent to .05 degrees C.

One of the outputs goes to a bi-colored led which will tell if the unit is heating or cooling. Another output goes to a 16 amp isolated zero crossing triac which turns on the power to a device connected to it.

A dip switch provides 16 different temperatures to which the controller can be pre-programmed to. The thermistor can be used from -40°C to 150°C.

BUILDING THE TEMPERATURE CONTROLLER BOARD
By Ron Newton
Revised April 20, 2009

See parts list on CD for components.

1.   The components locations are marked on the board. Solder the 2 fuse clips (make sure the ears are on the outside), ZNR which goes next to the fuse clips, 2ea 220 ohm resistors, 1ea 10K resistor and the .1 uF capacitor to the board.

2. The 16F819 is preprogrammed and can be soldered to the board. It also can be reprogrammed using the programming pins and a Microchip PicKit 2 programmer. Make sure to put pin one to the square pad.

3. Solder the terminals, 5 programming pins and the dipswitch to the board. Pin one goes to the square pad.

4. Solder the 1000 uF cap, bridge diode and the voltage regulator to the board. Note the polarities. If using a LM7805 220 mount pin one is on the left pin 3 going to the chip. If using a LM 7805 92 the flat should be facing inwards toward the capacitor, pin one going to the chip.

5. Solder the led with the long lead going to the square pad. The bottom of the negative should be even with the bottom of the board. The led eventually will go through the hole in the lid.

6. If it is going to be used for 110, solder two jumpers on the board labeled 110. If it is going to be used for 220, solder the one jumper labeled 220 on the board. Solder the transformer to the board. Pin one is toward the fuse holder.

7. Clip the transformer leads and the terminal leads close to the board as the board sits close to the chassis. No terminal leads should touch the bottom while resting on the spacers.

8. Bend the leads of the triac toward the top of the triac where the leads change thickness up at a right angle. Solder the leads from the bottom of the board so that they rest on the bottom. The body of the triac should be half way on the side of the board. Smear heat sink compound on the bottom of the triac. Loosely mount the triac to the inside of the chassis using a 4-40 screw from the bottom and a nut.

9. Place 4 6-32 screws from the bottom and add 4 spacers. Screw the four screws into the board. The board will self thread.

10. Tighten the 4-40 screw and add a drop of fingernail polish to the nut.

11. Add the grommets to the chassis and the fuse.

12. To program, set all the dip switches on off.

13. Using Wire Wrap cut the length you need for the themistor to reach the item you are controlling. Solder the wires. Encapsulate the themistor by using a hot glue gun and empty plastic wire twist connector.

14. LM 7805 92 the flat should be facing inwards toward the capacitor, pin one going to the chip.

15. Solder the led with the long lead going to the square pad. The bottom of the negative should be even with the bottom of the board. The led eventually will go through the hole in the lid.

16. If it is going to be used for 110, solder two jumpers on the board labeled 110. If it is going to be used for 220, solder the one jumper labeled 220 on the board. Solder the transformer to the board. Pin one is toward the fuse holder.

17. Clip the transformer leads and the terminal leads close to the board as the board sits close to the chassis. No terminal leads should touch the bottom while resting on the spacers.

18. Bend the leads of the triac toward the top of the triac where the leads change thickness up at a right angle. Solder the leads from the bottom of the board so that they rest on the bottom. The body of the triac should be half way on the side of the board. Smear heat sink compound on the bottom of the triac. Loosely mount the triac to the inside of the chassis using a 4-40 screw from the bottom and a nut.

19. Place 4 6-32 screws from the bottom and add 4 spacers. Screw the four screws into the board. The board will self thread.

20. Tighten the 4-40 screw and add a drop of fingernail polish to the nut.

21. Add the grommets to the chassis and the fuse.

22. To program, set all the dip switches on off.

23. Using Wire Wrap cut the length you need for the themistor to reach the item you are controlling. Solder the wires. Encapsulate the themistor by using a hot glue gun and empty plastic wire twist connector.

You may have to bypass the temperature control unit and substitute the Triac. Or you can set the refrigerator or freezer to the coldest setting (incubator to full on) and use the plug connecting to the Triac.

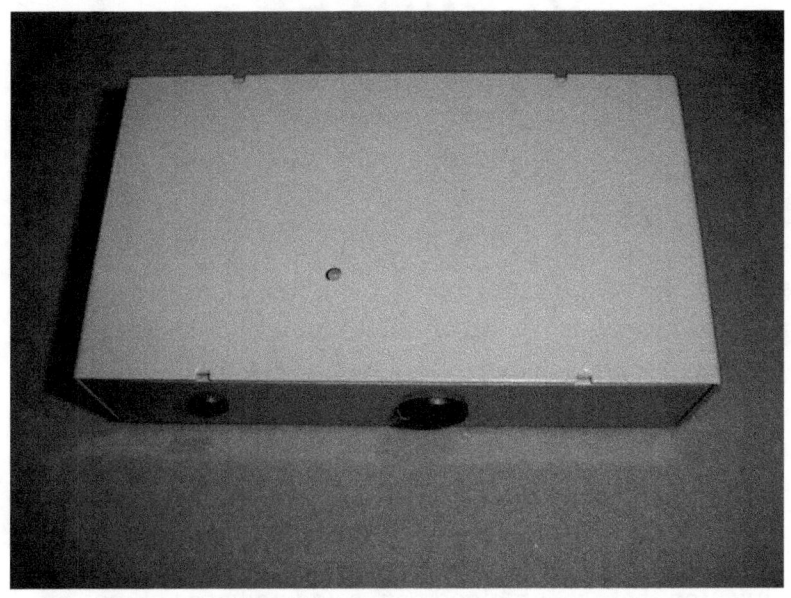

PRECISION TEMPERATURE CONTROLLER

FREEZER, REFRIGERATOR & INCUBATORS

CHAPTER SIXTEEN

HANDHELD BATTERY OPERATED SPECTROPHOTOMETER

Ron Newton
October 20, 2012

This project is of moderate complexity and the electronics are straightforward. A PIC microprocessor is used and requires a programmer for programming. The cuvette holder is best built using a milling machine and lathe. However, it can be built with only a drill press and a file. The cost of the project is under $50.00 and can be built in about 6 hrs.

As a retired Clinical Scientist, I donate my time to a non-profit organization that sets up blood banks and pediatric clinics in exotic places e.g. Honduras, Republics of Georgia, Armenia and Uzbekistan. We are now preparing for the launching of laboratories in 14 different areas in Africa and Haiti. I perform blood banking, water analysis, and medical laboratory analysis. One of the basic tools for any chemist is a spectrophotometer for measuring the intensity of colors of liquid solutions. With the right chemicals, the concentration of a compound can be determined by the amount of color it produces: E.g., free chlorine. If you have a swimming pool or a spa, you probably have measured the free chlorine by using a color comparator and adding a few drop of ortho tolidine (OTO). The intensity of the yellow color produced by the chemical reaction is proportional to the amount of free chlorine. Most water systems in the United States still use chlorine for the disinfection of bacteria. The problem is the amount is 1/10 of that of swimming pools and is difficult to determine by one's eye. The spectrophotometer solves this problem. Comparing drinking water with a known amount (standard) of chlorine is easily measured. There are hundreds of compounds that can be measured this way and there are volumes of books written on the subject. For the serious amateur scientist, a spectrophotometer is a must.

Spectrophotometers provide a light source with a given wavelength. The expensive ones use gratings and filters to insure the proper wavelength. They are heavy and cannot be jarred. The first models used a lamp passing the light though a gelatin filter to provide an approximate band of light. The photo detector was a 2 x 2-inch

$$OPTICAL\ DENSITY = \frac{-\ LOG\ \%\ TRANSMISSION}{100}$$

detector made out of selenium.   A mirror connected to a galvanometer reflected a light beam on the wall. The wall chart had two scales; one indicating the percent transmittance and another that was called optical density (OD) also known as absorbance.

If you have a standard amount of chemical and measure it's OD, you can obtain a factor by dividing the amount by the OD. If the OD of the unknown is known, by multiplying the OD times the factor you will get the amount of the unknown chemical. The circular slide rule was the ideal tool for this calculation in my day. Modern electronics has come to the rescue!

To digress, one of the problems in the field or in third world countries is the lack of power or an improper power source. I needed something portable, light weight, rugged and self-powered. I have been using this device for years and found it invaluable whether measuring free chlorine, hemoglobin, blood sugar, BUNs, etc.

129

ELECTRONICS:

This unit uses three colored leds in one convenient package that provides blue, red, and green wavelengths. They have sharp cutoffs of 470-nm, 534-nm and 636-nm respectively. These three wavelengths are capable of measuring hundreds of different compounds.

The power is supplied by a 9 volt battery which is reduced to 5 volts using a 7805 (IC4). C1 filters any ac out. A Microchip PIC16F619 18 pin microprocessor is used as it can be programmed on the board by using the programming pins on the bottom. It has several 1024 bit a/d (analog to digital) converts and only one is used. A three volt voltage reference is used on pin 5 the voltage reference pin. This helps in the amplification of the input signal. Several programming buttons are used and a six dip 10K resistor network keeps their inputs high until the buttons are pushed which puts them to ground. The microprocessor also controls the Dog series LCD display module which has two lines and can write 256 characters and numbers. This is a very versatile chip and only costs about $12.00. The micro also turns on the three leds providing the most common wavelengths in spectrocroscopy.

A pin photodiode is used for the detection of the amount of light transmitted through the cuvette. It is amplified by a MCP601 instrumentation amplifier (IC2). A programmable resistor is used (IC3) to zero the output which is 100% transmission or 0.00 optical density. When the blank button is pushed the programmable resistor starts at a high resistance and is decreased until 100% transmittance is detected (3 volt output from the photodiode) and then locked.

A 1024 look up table is used for the calculation of O.D. When in the concentration mode, the O.D. is divided by a variable factor that is set using the up and down buttons. The decimal button moves the decimal.

130

One of the problems encountered was that pin diodes have a dc dark current which prevents the amplifier from going to zero. This is compensated using a math algorithm subtracting out the dark current.

CUVETTE HOLDER:

A 1" nylon round 2.75" in length is used. A center hole is drilled using a 31/64" drill to the depth of 2". Drill and tap the center of the bottom using 43 drill and 4/40 tap. (See CD for plans.)

Using a # 8 drill, drill the Pin Diode. There is an offset due to the sensor area not being in the center. The distance from the bottom is 1.2755". Drill only to a depth of .2". Change the drill bit to 3/32" and the distance to 1.25" from the bottom. You may want to start the hole using a drill center or small mill as it will tend to drift to the center of the Pin Diode hole. Drill all the way through.

Rotate the holder 180 degrees and using the 3/32" drill makes sure that it is vertical. Change back to a # 8 drill bit and drill to a depth of .2".

Wire wrap the Pin Diode with about 6" of red wire to the anode and 4" of white wire to the cathode. Twist the wire together. Solder and clip off the excess leads. The flat should face the bottom. Push the Pin Diode into its hole and secure from the outside with black silicon glue.

Wire wrap the led with 6" white wire to the two cathodes. Wire wrap 6" red to the red anode, 6" green wire to the green anode and a 6" blue wire to one of the blue anodes. Solder and cut off the excess leads. Twist the wire together. Use two batteries with a 220 ohm resistor to insure that you have the right colors. Insert the led into its hole and put one drop of super glue to secure it in the holder. Use black silicon glue and coat it. Both the Pin Diode and the led must be light proof. (Both data sheets are located on the CD.)

BOX:

Put the box together with its 4 screws. On the CD is a template for drilling holes in the top and bottom. Cut the template out and using glue stick glue to the top of the box. The 1" hole should not be over the battery box. Using a #43 drill, drill the 1" hole in the center and through the bottom. This will make sure that the cuvette holder will be aligned with the top. Use a 1" hole saw and drill slowly through the top to prevent cracking. Drill the four corner holes with 7/16" drill. Drill the six button holes with a 9/64" drill and the power switch hole with at 9/32" hole. Remove the template with hot water

BOARD ASSEMBLY:

The board files for making the board are located on the CD. Go to www.expresspcb.com and download their free software to open the files. The top side of the board has the printing on top. Solder IC1, IC2 and IC3 chips to the board. Pin 1 goes to the square pads. Solder the display to the board and remove the protective tape. Solder all the switches. IC4's flat should face the off/on switch. D1's flat faces the micro. Solder both. Solder the four resistors and the two capacitors. Solder the resistor network with pin one to the square pad.

Feed the wires coming from the cuvette holder from under the board and solder. The board is marked. The anode from the pin diode goes to the pad connected to pad three of the micro and the cathode to pin 2. The cathode from the led is connected to R10.

Open the battery compartment and thread the 9 volts wire through the right slot with the battery holder on the bottom looking from the top. Feed the two wires through the strain relief hole and solder the red to the + and black to the remaining pad next to it.

FINAL ASSEMBLY:

Mount the cuvette holder in the bottom with a 1/4" 4-40 screw. Using 3/8" nylon spacer and 4 6-32 screws mount the board to the top of the box.

Place a 9 volt battery in the battery compartment. Press the on/off switch. The LCD should display" % TRANSMITTANCE" on the first line and number and "Red" after the number.

Hold the "Blank" button down until it reads about 100.0. Pressing the "Mode" button, the display should change to "Optical Density" pushing again to "Concentration". Pushing the "Wavelength "button the display should change from "Red" to "Green" to "Blue".

When using a cuvette, make sure to put the cap on to prevent stray light.

A SIMPLE PRACTICAL USE:

Using a fresh bottle of bleach or sodium hypochlorite also known as Clorox add 3.3 mls[1] of bleach to 1000 mls of distilled water. Mix well. Using 5 mls of this solution dilute to 1,000 mls again. This will make an approximate standard of .500 parts per million (PPM) or 0.5 mg/l of free chlorine. It needs to be made up fresh each time. Place 5 mls of the standard into a cuvette and add 5 drops of ortho tolidine (OTO). Mix by inverting. Take 5 mls of tap water and add 5 drops of OTO and mix by inverting.

1.   Turn on the meter and press the wavelength button until the blue displays.
2.   Press the mode button until the Concentration display.
3.   Push the decimal button until the decimal[2] reads 00.00

---

[1] Normal bleach contains approximately 30,000 – 50,000 ppm of free chlorine.
[2] The decimal button does nothing to the calculations but only acts as a reference point.

4.   Place the blank into the well, cover and push the blank button. The display should show 0.
5.   Place the standard into the well, and adjust the concentration using the up and down buttons until the display shows 0.500.
6.   Place the tap water into the well and read the concentration.

Normal tap water usually contains from .01 to .5 PPM free chlorine.

WHY OPTICAL DENSITY?

Spectroscopy came in to being in the early 20th century. The photodiode produces a linear response in voltage to the amount of light striking it. However, if you recall from high school physics, light follows the inverse square law. That is to say, double the distance and you receive only the ¼ the amount of light. The same thing applies to colored solutions. Double the concentration and you only get ¼ of light. If you plot concentration vs. % transmission on regular graph paper, you will get a curve. The first methods use semi-log paper and plotted the concentration using three points varying the concentration. If a straight line occurred it was said "It followed Beer's law". Nothing to do about the making of beer, he was one of the people to discover the relationships.

Converting the % transmission to optical density (-log %/100) straightens up the curve and now it can be plotted on regular graph paper. The concentration mode of any spectrophotometer assumes that the reaction follows Beer's law, but not all reactions do! This is why optical density is on the spectrophotometer. If the reaction does not follow Beer's law, you need to run several standards of different concentrations plot the points on regular graph paper and connect the dots obtaining a curve. The unknown OD is then found on the curve and its concentration is then determined.

Optical density is also used in kinetic reactions at a given temperature. Enzymes can be measured using optical density and

$$UNITS = \frac{\Delta \; OPTICAL \; DENSITY \; FOR \; T \; MIN}{T}$$

stop watch. The first OD is noted and a stopwatch is started. After a period of time (hours, minutes or even seconds) the change of OD is noted.

Many cardiac enzymes are measured using this method.

Happy Spectroscopy!

THREE WAVELENGTH
SPECTROPHOTOMETER HAND HELD

CHAPTER SEVENTEEN

LET'S COUNT GEIGERS
A RECORDING RADIATION
COUNTER

Ron Newton
December 24, 2012

In 1990, my wife and I had the great opportunity to visit and live in Mother Russia during the Communist era. We stayed with a family in Kiev located in the Ukraine. The Chernobyl disaster had occurred 4 years earlier and they were still washing down the streets daily. Perhaps a harbinger of what may happen in the USA. In those days, it was illegal in Russia for the people to possess a Geiger counter and they had to rely on local authorities. When Chernobyl happened, the people of Kiev were not informed officially and only heard about it through the grapevine.

Fedor our host, (also a Russian Physicist) was most concerned about his grandchildren and the effects of the radiation. Rumors were still flying after 4 years and no one really knew the facts. Milk and produce were their main concern. He asked me if I could provide him with a Geiger counter. When I returned to the States I decided to put one together and sent it to him. The radiation levels he found were of concern, especially in tree fruit.

Now the USA is faced with a similar situation. It is not a question of "If" only "When" a dirty bomb will be set off by terrorists. It's not that I distrust the bureaucrats, but I like to gather my own facts, and make sure the correct information is being disseminated to the public. I recently re-examined my schematics of the unit I previously built and added a microprocessor for control and a new display. I recommend a LND 712 Geiger tube which appears to be the Gold Standard of portable counters. It is capable of measuring alpha, beta, and gamma particles. If you don't want to spend a lot of money on a Geiger tube, you can find some on E-bay for under $5.00. You may need to adjust the high voltage for the tube. This is discussed in "Hints and Tips" on the article link. The cost of the project is less than $35.00 in parts excluding board and tube

This is great project for high school and University students. It is simple to assemble and teaches a number of techniques in building of projects. This unit detects and accurately displays levels of radiation. The unit can detect and display dosage level as low as one micro-roentgen/hr to as high as micro-roentgens/hr. It can also detect Radon which emits alpha particles. This unit has the advantage over other kits as it has an output for recording months of data showing trends of radiation with a data logger.

ELECTRONICS:

The brain of the counter is a PIC16F916 microprocessor. You will need a programmer to program it. The chip is preprogrammed in the kit.

Geiger tubes require high voltages often over 500 volts; however, they pull a very small amount of current in the micro-amps. The high voltage is generated by Q1 and Q2 and the choke. Q3 reduces the amount of power the unit uses thus increasing battery life. A combination of high voltage capacitors and diodes increase the voltage as they are in a voltage tripler configuration. A zener diode controls the voltage to the tripler by a feedback to Q1. I used a 270 volt surface mount zener. In theory, tripled should produce 810 volts but in practice it produced 600. Keep in mind that the circuit generates very little current therefore trying to measure it with a standard DVM will load down the voltage reading. However, touching any part of the high voltage circuit will give a startling effect.

The output from the tube (note it comes from the case of the tube) is converted to a suitable voltage using Q4 npn transistor to the microprocessor. The micro has a built in timer and is capable of counting seconds, minutes, hours and days. A momentary switch programs the time for collection which is displayed on the LCD.

The numbers of counts per minute are converted by the digital to analog converter (DAC) and this voltage is available via the RCA jack.

The unit is housed in a Serpac case with a 9-volt battery insert.

BUILDING THE UNIT:

The board files from ExpressPCB.com are located on article's link along with assembly files for the microprocessor. The microprocessor can be programmed on the board using a PIC 2 programmer.

Solder IC1 and IC2 to the top of the board noting pin one has the square pad. D5 is the surface mounted zener and has an indented line across one end. This is the cathode. Solder D5 to its pads. You will see another set of pads to the right of D5 with a trace across shorting the pads. Do **NOT** solder D5 to these pads; they are for an extra zener if using a different Geiger tube other than a LND 712. Solder the remaining diodes noting their polarity. Solder IC3 noting its flat pointing to the switch. Solder in the transistors. Solder C4, C5, C8 and C6 noting its polarity, and all the resistors. Solder the two switches and the LCD.

Now turn the board over and place C1, C2 and C3 on the bottom side and solder. C7 also goes on the bottom side. Lift it off the board about 2-3 mms so you can solder L1; note its polarity and solder.

Place L1 on the top of the board and solder.

CHASSIS:

Using the four screws, screw the top and the bottom of the chassis together. The screws will self tap and it will make it easier for final assembly and secure the box for drilling.

Go to the CD and down load the template. Cut out the templates and using glue stick glue the top template to the top and the template to the sides of the box. The battery box should be on the top. Drill the holes as marked and deburr the holes. Hot water will remove the templates.

Place the 9 volt battery snap inside the chassis (facing the bottom) and run it through the slot on the right side. Run the battery wires through the strain hole of the board and solder the red wire to the + and black wire to the -. Solder the wires.

Using two 3/8" standoffs, mount the board to the top chassis using 1/2" 6-32 screws. The holes in the board will self thread.

The end window of the LND 712 is very thin and fragile and it is easy to poke a hole into. (I speak from experience.) Strip a 4" piece of #30 blue wire wrap on the short wire coming from the side of the tube. Solder a 4" #30 wire wrap red wire to end clip of the Geiger tube. Place a piece of insulated tubing over the end clip to prevent shock. Coil both the wires. Mount the Geiger tube next to its hole using a hot glue gun. Solder the red wire to HV+ pad and the blue wire to IN pad.

Wire wrap a 4" red wire to + post of the speaker and a 4" blue wire to the other post. Coil the wires. Glue the speaker to the side of the box with hot glue. Solder the blue wire to the round pad and the red wire to the square pad on the pads marked SPK.

Place the RCA jack into its hole. Solder 4" of red and blue wires to the RCA jack. Red goes to the center pin. Solder the red wire to the square pad (pin 8 IC2) and the blue wire to round pad (pin 7 IC2). Coil the wires using a 1/8" in diameter screw driver or 1/8" drill and coil the wires for easy storage.

Secure the box with the 4 screws.

TESTING THE UNIT:

Snap a 9-volt battery into the battery clip. Push the on-off switch. The display should show readout in CPM (counts per minute) which is the default. As radiation is detected, a click will be heard and the counter will increase. Pushing the mode button will advance the display from rems per minute, hours and days. If the mode in held down for more than 5 seconds it will reset all the counters and registers. If in micro-rems per minute, the unit will count 60 seconds and then latch the counter showing the count for the minute. The unit will still be counting, demonstrated by the click and the flashing of a t on the display, but will not display a running count until the next minute. A count down in seconds will be shown until the results toggle. The same occurs for hours and days. The unit will automatically changes from micro-rems to milli-rems as needed.

The unit will alarm for five seconds if the count exceeds a calculated .5 micro-rems in a minute period (30 µR/Hr). To turn off the alarm and beeping, turn the unit off and hold down the mode button and turn the unit on again.

The background count will vary with the area you live in. 12 µR/Hr is about average. I get 20 per hour at my laboratory. This is probably due to the altitude of Carson City. Before 9-11, I took it in an airplane and turned it on at 35,000 feet. It will make you wonder why you are flying. 300 µ/R Hr. Pick up a Coleman lantern mantel and use this to test the unit. It contains thorium that is radioactive and emits alpha particles. The unit draws about 10 ma and a 9 volt battery will last about 2 days. Use a 9 volt dc battery eliminator if recording for days or months.

MAKING A RADIATION GRAPH:

I use two types of devices for making graphs. The first is a Dataq EL-USB-3 which a voltage data logger that plugs into the USB port of computer. This device makes the Geiger counter truly portable. It sells for about $75.00. It measures 0 - 30 volts and can be set to record from one second to 12 hour interval. At one minute interval, it will record 22 days of data. It is only sensitive to 50 mV which is 41 CPM.

The second device I use is my favorite. It is the Dataq DI-145 which sells for $29.00. It is a must for any scientist. Most of us have an older computer sitting around gathering dust. This device will record up to four channels at one time and with Dataq's free WinDaq's software is very powerful tool. You could monitor gravity, tilt of the earth plus seismic activity, magnetic fluxuations of the earth and radiation all at the same time and see if there are any correlations. The software graphs out all on the same sheet. Just plug in either device into the RCA jack. Its sensitivity is .001 volt which is 1 CPM.

WHAT ARE WE MEASURING?

Well as usual, when entering in the scientific world there is a plethora of terms and units. Fortunately for our purposes 1 roentgen approximately equals 1 rad equals one rem. The roentgen is the quantity of x-rays or gamma rays. One roentgen is a lot of radiation and is equivalent to about three to five years of normal exposure. This is why milli-roentgens and micro-roentgens are used. Milli-roentgen is equal to 1/1000 (1×10^{-3} roentgens) and a micro-roentgen is equal to 1/1,000,000 (1×10^{-6} roentgens). The unit we are building measures in micro-roentgens. Due to a number of factors (type of tube, calibration and different types of radiation) consider one click or one count 1/105 of a micro-roentgen (1×10^{-6} roentgens).

The rad (Radiation Absorbed Dose) is the unit for measuring absorbed doses of radiation.

The Rem (Roentgen Equivalent Man) is the damage that one roentgen causes to man.

There are three main types of ionizing radiation. (Ionizing radiation causes changes in tissues and can cause cancer.)

142

Alpha particles are positive charges that are very short lived. A sheet of paper can stop them, however if ingested or inhaled they become very deadly. Radium is naturally radioactive element. When it emits an alpha particle it becomes radon, a radioactive gas with a half-life of 3.8 days. Just as there are earth tides, (see Nuts & Volts May 2003) it has been reported radon tides also vary day from day.

Beta particles are much smaller than alpha, are negatively charged but can be blocked by a thin sheet of aluminum foil.

Gamma rays are electromagnetic radiation and are stopped by lead, concrete or steel.
Depending on what you read, we are exposed to about 100 – 200 milli-roentgen per year. A chest x-ray (low level gamma rays) gives us about 10-milli-roentgen exposure.

When to become alarmed? This is difficult to say, however the EPA forbids radiation exposure above 5 rems in any one year to radiation workers.

Said, Winnie the Pooh to Christopher Robin, "It is interesting that there are Geiger counters but no Giegers to count, only Tiggers!"

RECORDING GEIGER COUNTER

www.ingramcontent.com/pod-product-compliance
Lightning Source LLC
Chambersburg PA
CBHW051528170526
45165CB00002B/657